CHEMISTRY RESEARCH AND APPLICATIONS

SOLUTION CHEMISTRY

ADVANCES IN RESEARCH AND APPLICATIONS

CHEMISTRY RESEARCH AND APPLICATIONS

Additional books in this series can be found on Nova's website
under the Series tab.

Additional e-books in this series can be found on Nova's website
under the eBooks tab.

CHEMISTRY RESEARCH AND APPLICATIONS

SOLUTION CHEMISTRY

ADVANCES IN RESEARCH AND APPLICATIONS

YONGLIANG XIONG
EDITOR

nova
science publishers
New York

Library of Congress Cataloging-in-Publication Data

ISBN: 978-1-53613-101-7

Published by Nova Science Publishers, Inc. † New York

CONTENTS

Preface		**vii**
Acknowledgments		**xi**
Chapter 1	Experimental Study and Thermodynamic Modeling of Niobium, Tantalum, and Uranium Behaviour in Supercritical Fluoride Hydrothermal Solutions *Alexander F. Redkin, Nataliya P. Kotova, Yury B. Shapovalov and Vassily I. Velichkin*	**1**
Chapter 2	Experimental Determination of Brucite Solubility in NaCl Solutions at Elevated Temperatures *Leslie Kirkes and Yongliang Xiong*	**47**
Chapter 3	Confined Water in Rock Salt *Mei Ding, Bernard H. W. S. de Jong, Erik Watkins, Monika Hartl, Luc Daemen, Hylke. N. de Jong and Qingcheng He*	**69**

Chapter 4	Absence of Mineral Colloids in High Ionic Strength Solutions Associated with Salt Formations: Experimental Determination and Applications for Nuclear Waste Management *Yongliang Xiong, Leslie Kirkes, Sungtae Kim, Cassie Marrs, Jandi Knox, Justin Dean, Haoran Deng and Martin Nemer*	**97**
Chapter 5	Conformational Restructuration of Ionic Polysulfones in Solution: Perspectives towards Smart Biomaterials *Anca Filimon*	**133**
Chapter 6	Determination of Trace Elements in Crude Oils and Fuel Oils: A Comprehensive Review and New Data *Weihang Yang, Yongjun Gao and John F. Casey*	**159**
Chapter 7	Solution Chemistry of Transition Metal Complexes: A Case Study of Vanadium *Daniele Sanna*	**207**
Chapter 8	Modeling Actinide Solubilities in Alkaline to Hyperalkaline Solutions: Solubility of $Am(OH)_3(s)$ in KOH Solutions *Yongliang Xiong*	**229**
About the Editor		**241**
Index		**243**

PREFACE

Examples of solution chemistry are ubiquitous in our society and in nature. Solution chemistry is a key component in industrial processes, and natural processes ranging from geochemical processes related to weathering reactions at the Earth's surface to the formation of hydrothermal ore deposits from magmatic/meteoric fluids, and to the potential migration of radionuclides from geological repositories to the geosphere. The aim of this book is to present original data and comprehensive reviews of state-of-the-art work that advance our understanding of solution chemistry in diverse fields, authored by the diverse experts.

Chapters 1 and 2 address advances in solubility measurements. The first chapter provides the experimental data and thermodynamic modeling regarding solubilities of uraninite [UO_2(cr)], pyrochlore [(NaCa) Nb_2O_6F(cr)] and microlite [(NaCa)Ta_2O_6F(cr)] in NaF solutions under the extreme conditions at 800°C and 2,000–2,300 bars. The knowledge embodied in the first chapter is important for understanding the formation of rare metal deposits in the ultrahigh temperature hydrothermal systems. The second chapter presents an experimental study of brucite [$Mg(OH)_2$(cr)] solubilities in NaCl solutions over a wide range of ionic strength from 0.01 mol•kg^{-1} to 5.6 mol•kg^{-1} at elevated temperatures up to 80°C. Such results are useful in many areas of study, including elucidation of the formation of recently discovered brucite-carbonate chimneys in hyperalkaline (pH 9–11), saline, hydrothermal fluids (40°C to 90°C) on ocean floors.

Chapters 3 and 4 address advances in solution chemistry related to salt formations. In Chapter 3, the incoherent inelastic neutron scattering (INNS) technique was used to distinguish three water environments in bedded salts. These three environments include intergranular water molecules confined to grain boundaries, water in fluid inclusions, and structural water in hydrous minerals. Such knowledge is relevant to subsurface energy exploration and nuclear waste storage in salt formations. Chapter 4 provides the experimental evidence to demonstrate that mineral colloids derived from an engineered barrier of MgO, in high ionic strength brines associated with salt formations are nonexistent. This is attributed to the fact that the high ionic strength solutions associated with salt formations have high concentrations of both monovalent and divalent metal ions that are orders of magnitude higher than the critical coagulation concentrations for magnesium oxide derived colloids. The absence of magnesium oxide derived colloids in high ionic strength solutions implies that contributions from these colloids to the total mobile source term of radionuclides are minimal in a salt geological repository.

Two chapters summarize advances related to organic materials. Chapter 5 provides an overview regarding the advances in new functionalized polysulfones. Recently developed polysulfonic films are expected to find biomedical applications.

Chapter 6 includes a comprehensive review with regard to the analytical techniques for determination of trace elements in crude oils and fuel oils, and the original data generated in the laboratory. Based on these results, it seems that elements of high value could be co-extracted from crude oils in the future.

The final two chapters address advances in speciation in understanding in aqueous systems. Chapter 7 presents an overview of the advances in speciation of vanadium using three techniques: Nuclear Magnetic Resonance (NMR), Electron Paramagnetic Resonance (EPR), and UV-Vis spectroscopy.

Chapter 8 centers on the speciation of Am(III) in hyperalkaline solutions. The equilibrium constant for $Am(OH)_4^-$ is obtained by using the Pitzer model to evaluate $Am(OH)_3(s)$ solubility data in KOH solutions.

ACKNOWLEDGMENTS

Sandia National Laboratories is a multimission laboratory managed and operated by National Technology and Engineering Solutions of Sandia, LLC., a wholly owned subsidiary of Honeywell International, Inc., for the U.S. Department of Energy's National Nuclear Security Administration under contract DE-NA-0003525. Dr. Dien Li at Savannah River National Laboratory, and many other referees who wish to remain anonymous are thanked for their reviews. SAND2017-11768J

Yongliang Xiong,
October 16, 2017
Carlsbad, NM, US

In: Solution Chemistry
Editor: Yongliang Xiong

ISBN: 978-1-53613-101-7

Chapter 1

EXPERIMENTAL STUDY AND THERMODYNAMIC MODELING OF NIOBIUM, TANTALUM, AND URANIUM BEHAVIOUR IN SUPERCRITICAL FLUORIDE HYDROTHERMAL SOLUTIONS

Alexander F. Redkin[1,*], Nataliya P. Kotova[1], Yury B. Shapovalov[1] and Vassily I. Velichkin[2]

[1]Laboratory of Model of Ore Deposit,
Institute of Experimental Mineralogy Russian Academy of Sciences,
Chernogolovka, Moscow Region, Russia

[2]Laboratory of Radiogeology and Radiogeoecology,
Institute of Geology of Ore Deposits, Petrography, Mineralogy, and
Geochemistry Russian Academy of Sciences, Moscow, Russia

ABSTRACT

The NaF effect on uraninite, pyrochlror and microlite solubility at 800°C is studied. The immiscibility boundaries and compositions of fluid

[*] Corresponding Author Email: redkin@iem.ac.ru.

phases L_l and L_2 are defined in the system NaF–H_2O. It is shown that the solubility of the minerals being studied is higher in the phase L_l than in L_2. It is established that microlite solubility even in highly concentrated fluoride solutions is very low. It is found that the interaction of the fluoride solution with UO_2 in the immiscibility region leads to the formation of uranium fluoride $NaUF_5$ H_2O. The model of "acidification" L_l and "alkalizing" L_2 is supposed.

Keywords: pyrochlore, microlite, uraninite, fluid miscibility, fluorite solutions, solubility, thermodynamic modeling, species

1. Introduction

The migration forms of ore components, the mechanisms of their accumulation and separation are of great importance for the construction of quantitative models of the deposits formation. Of particular interest are the rare metals tantalum and niobium, whose compounds have low solubility in dilute aqueous solutions and the possibility of their transfer by hydrothermal solutions, it seems unlikely. Although, the contrasting deposition of rare-metal ores confined to carbonatite rocks can be considered as a process of deposition of niobium and tantalum in carbonate geochemical barrier. As for the rare metal ore formation in granite rocks, in Orlovka, Etyka tantalum deposits and Achikan ore occurrence located in Hangilai rare metal ore field in the south-western spurs of Bortevoch Ridge in the Eastern Transbaikalia, the question of the mechanism of tantalum accumulation is still a matter of debate. It is known that the formation of rare-metal ores in Li-F granites in these deposits was quite long and proceeded in several stages. Generally accepted are the genetic representations of V.I. Kovalenko (Kovalenko 1977), consisting in that the rare-metal Li-F granites accommodating tantalum ores are formed by the most deeply advanced fractional crystallization of ordinary granite magma under specific conditions, providing stepwise accumulation in the residual granite melt of F, Li, Ta, Nb, U and other rare metals. These ideas were developed in the geochemical studies of A.M. Aksyuk (Aksyuk 2009, Zaraisky 2009), who showed that the indicator of the differentiation depth of tantalum-bearing granites of the Kukulbei intrusive complex is the Zr/Hf ratio, and solutions that are in equilibrium with Li-F granites contained high concentrations of fluorides. According to ideas of G.P. Zaraisky (Zaraisky 2008), a magmatic water-fluoride fluid is not able to leach out all the niobium from the melt, but this fluid has a high dissolving power with respect to tantalum and niobium minerals, which makes possible subsequent mobilization and redeposition of these metals. It is obvious that a solution that is in equilibrium with the melt and deposited in it with ore minerals is not sufficient for concentrating tantalum ores. A huge role is assigned to the water-saturated granite magmas. The water content in these granite melts reaches 8 wt.%, i.e., in a ton of melt contains 80 kg of H_2O or its solution. According to the available data (Dolejš 2007, Redkin 2009), sodium salts (NaCl, NaF) are the main components of hydrothermal solutions, equilibrium with melts of Li-F granites. High concentrations of fluorides, especially in those solutions in which fluid immiscibility is observed, can dramatically change the solubility of ore minerals. Fluid immiscibility in the NaF-H_2O system, according to experimental data (Kotelnikova 2008) occurs at T> 750-780°C and a pressure of about 200 MPa. In connection with the above, most of the experiments carried out in aqueous solutions of sodium fluoride at 800°C and a pressure from 170 to 230 MPa.

2. Methods

2.1. Experimental and Analytical Techniques

Experiments on mineral solubility were performed at 800°C, P = 200 - 230 MPa using two types of apparatuses: hydrothermal high pressure apparatuses in reactors having an internal diameter of 6-7 mm made of a heat-resistant stainless nickel alloy EP-455; a gas pressure vessel.

The runs were carried out in sealed platinum capsules (0.4-0.5 cm diameter, 4-5 cm long, 0.01-0.02 cm wall thickness–under reducing conditions (buffer Co-CoO). A cylindrical leaky Ni-container with a mixture (200 mg) of metallic cobalt powder and cobalt oxide was placed in a reactor under Pt capsules. After the run, the Co-CoO buffer was checked by XRD analysis to ensure that both phases were still present. Reactors of the hydrothermal installation, in the absence of oxygen buffer mixtures, produced a partial oxygen pressure in the hydrogen permeable capsules corresponding to the Ni-NiO (NNO) buffer, whereas in the gas vessel, where the pressure was set by argon, $\log f(O_2)$ GB = $\log f O_2$(NNO) + 0.25 MPa. The run duration was 7-9 days at 800°C.

Into the Pt capsule were put 10–15 mg of Mic, 0–5 mg of Fl, 0–100 mg of NaF, and of 0.15 ml of H_2O. The content of the capsules was leached with 5 ml of 0.01 m HCl solution of high purity into conical tubes. The solution was then separated from solid phases by centrifugation at 6000 rpm. An aliquot (4 cm^3) of pure (without mineral suspension) solution was acidified with 0.2 cm^3 strong (~65 wt.%) ultrapure nitric acid. The obtained solution was analyzed using the ICP-AES and ICP-MS methods. The solid phases were studied by the XRD and REM (CamScan, Vega-II) methods. According to analyses, microlite and very small amounts of fluorite and villiaumite were present in the solid products. The villiaumite was usually present in run products conducted in highly concentrated (>15 wt%) solutions of NaF.

Minerals used for the experiments were microlite (Mic) (NaCa)Ta_2O_6F (Figure 1-a) and pyrochlore (Pcl) (NaCa)Nb_2O_6F (Figure 1-b), obtained by hydrothermal synthesis (Redkin 2013$_1$) from equimolecular mixtures $NaF{\cdot}CaCO_3{\cdot}Ta_2O_5$ and $NaF{\cdot}CaCO_3{\cdot}Nb_2O_5$ at 800°C, 200 MPa in 1 m NaF during 7 days; uraninite (Urn) $UO_{2.03}$ (Figure 1-c) obtained by hydrothermal synthesis from chemical U_3O_8 at 500°C, 100 MPa, $f(O_2)$ = $10^{-14.7}$ Pa (NNO buffer) in solution of 0.1 m HCl for 14 days. Fluorite (Fl) CaF_2 (Figure 1-d) was prepared by recrystallization of the ultrapure reagent CaF_2 in 0.1 m HF at 500°C and 100 MPa within 2 weeks. Then it was dried at 100°C and annealed in air at 500°C. The chemicals Nb_2O_5, NaF, CaF_2 HCl, HF used in the experiments were high reagent grade, whereas Ta_2O_5, U_3O_8 were p.a. (pro analysis) grade. All solutions were prepared with double distilled water.

Figure 1. SEM images of the typical synthetic microlite (**a**), pyrochlore (**b**), uraninite (**c**) and fluorite (**d**) used in the experiments on solubility at 800°C, 200 MPa.

2.2. A Conceptual Model of Minerals Dissolution

The investigations included two homogeneous regions (HS-I-homogeneous solution of "low" NaF concentrations and HS-II-high concentrated NaF solution), as well as the immiscibility field of fluid phases L_1 (fluid of moderate density) and L_2 (dense fluid), situated between HS-I and HS-II. The concentration of ore element (Me) at constant *T-P* parameters in homogeneous solutions HS-I and HS-II, saturated with ore mineral should be described by the continuous function:

$$m_{Me} = \sum_{i} n \cdot m_i, \tag{1}$$

$$m_i = F(K_p, m_{NaF}^k, X_{H2O}, m_{HF}^v)$$

where m_i - molal concentration of Me in the form of *i* specie which is a complex function of equilibrium constant of the reactions (K_p), of molal concentration of NaF, HF and mole fraction of water in corresponding degrees (*k*, *v*).

Apparent solubility of ore mineral and Me weight concentration change linearly with NaF weight concentration in the field of fluid immiscibility (from L_1 to L_2). Apparent solubility corresponds to the total concentration of component obtained at the association of two and more fluid phases in corresponding ratios into a single homogeneous fluid for example under cooling. According to the lever rule,

$$C_{Me} = X_{L1} \times C_{Me1} + X_{L2} \times C_{Me2} \tag{2}$$

where X_{L1} и X_{L2} - mass fractions of L_1 and L_2 fluid phases, C_{Me1} and C_{Me2} – wt.% of Me in L_1 and L_2 phases, respectively. After some transformations (see (Redkin 2015_1) equation 2 assumes the form of a linear function on mass concentration of NaF (C_{NaF}):

$$C_{Me} = A \times C_{NaF} \times (C_{Me1} - C_{Me2}) + B \times (C_{Me1} - C_{Me2}) + C_{Me2} \tag{3}$$

where $A = 1/(C_{L2} - C_{L1})$ and $B = C_{L2}/(C_{L2} - C_{L1})$ – constants which only depend on mass concentrations of NaF in L_1 (C_{L1}) $_a$ and L_2 (C_{L2}) phases, respectively. It should be noted that equation 3 correctly describes apparent solubility of a one-component ore mineral, as well as incongruent solubility of complex minerals in which chemical potentials of all the other components except for the main ore component have constant value.

Therefore, we chose the reactions of incongruent dissolution of microlite and pyrochlore with fluorite formation for our investigations.

2.3. Thermodynamic Approach and Evaluations

Experimental data on solubility of mineral ore phases were necessary to solve an inverse thermodynamic problem– the estimation of thermodynamic properties of niobium, tantalum, and uranium species in aqueous solution at 800°C and 200 MPa. Obviously, the effect of pressure on the results of the experiments is slight.

The task of finding and refinement of thermodynamic properties of aqueous solution species is solved using the software package HCh (Shvarov 1999) and the program supporting technology OLE Automation (Shvarov 2008), realized in the program OptimA (Shvarov 2015). The activity coefficients of the individual ions and species in solution are expressed in terms of Debye-Hűckel theory. A common expression is accepted in the form of extended D-H equation

$$log_{10}\gamma_i = -\frac{aZ_i^2 I^{0.5}}{1+4.5bI^{0.5}}$$

where I is ionic strength, Z_i is the ion charge, 4.5 is the size parameter for all charged species in angstroms (Rafalsky 1973), a and b are the dielectric water depended parameters that vary upon temperature (Helgeson 1974_1, Shock 1992). Ionic strength

$$I = 0.5 \times \Sigma(m_i \times Z_i^2)$$

is defined in terms of molal concentration. Changes in dielectric constant, structure, and Debye-Hűckel parameters of water at high temperature lead to enhancement by use extended D-H equation with progressively higher concentration (Helgeson 1974_2).

3. Experimental Results

3.1. Microlite Solubility

The results of the solution analyses after the runs on incongruent solubility of microlite at 800°C and 170, 200 and 230 MPa are shown in Figure 2-a and Tables 1-3. The data obtained show the microlite low solubility even in high concentrated fluoride solutions of Na. The trends of the change of m Ta on m NaF and mole fraction of water $X(H_2O)$ in the regions of homogeneous solutions of HS-I and HS-II at 800°C and 170, 200 and 230 MPa correspond to equations 4 and 5 in the range of 0.001 to 1.0 m NaF:

$$\text{m Ta} = 2.555 \times 10^{-7} + 2.284 \times 10^{-6} \times (\text{m NaF})^{0.5} - 1.293 \times 10^{-9} \times (\text{m NaF})^{-0.5} \quad (4)$$

in the range of 1.0 to 26.5 m NaF:

$$\text{m Ta} = 2.488 \times 10^{-6} + 2.2367 \times 10^{-6} \times \text{m NaF} + 1.0757 \times 10^{-7} \times (\text{m NaF})^2 + 2.439 \times 10^{-9} \times (\text{m NaF})^3 \quad (5)$$

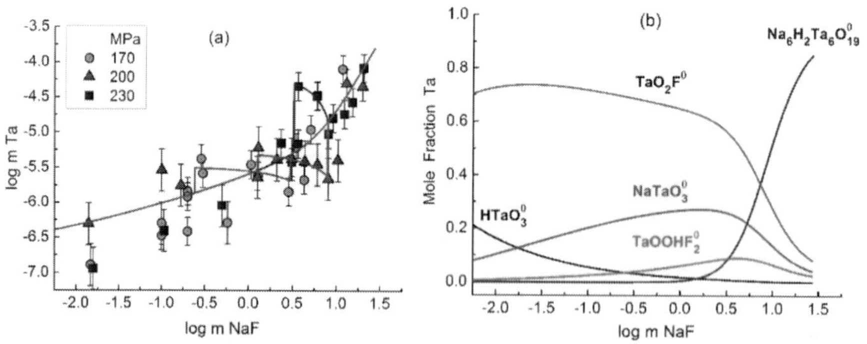

Figure 2. Effect of m NaF on microlite solubility (a) and tantalum speciation (b) at 800ºC, 170-230 MPa. (Experimental and calculated data.)

Table 1. Effect of m NaF on microlite solubility at 800°C and 170 MPa. Run duration 7–9 days

Initial solution			Quenched solution		Run
NaF wt.%	m NaF	log m NaF	log m Ca	log m Ta	no
0.0		-1.83*	-2.78	-6.89(±0.3)	123Kot
0.42	0.10	-1.00	-3.42	-6.30(±0.3)	124Kot
0.83	0.20	-0.70	-2.90	-5.84(±0.2)	133Kot
1.24	0.30	-0.52	-2.52	-5.58(±0.2)	134Kot
2.36	0.58	-0.24	-2.75	-6.29(±0.2)	125Kot
4.27	1.06	0.026	-2.45	-5.46(±0.2)	126Kot
8.08	2.09	0.104	-2.75	-5.62(±0.2)	130Kot
30.32	10.36	0.454	-2.83	-5.85(±0.2)	131Kot
35.24	12.96	0.63	-2.66	-5.68(±0.2)	132Kot
17.56	5.07	0.705	-2.80	-4.96(±0.2)	127Kot
32.86	11.66	1.07	-2.79	-4.10(±0.2)	129Kot
0.83	0.20	-0.70	-2.38	-5.92(±0.2)	145RAF
0.42	0.10	-1.00	-3.15	-6.48(±0.2)	146RAF
0.83	0.20	-0.70	-3.15	-6.42(±0.2)	147RAF
1.19	0.29	-0.542	-2.88	-5.38(±0.2)	148RAF
12.64	3.45	0.537	-2.82	-5.21(±0.2)	149RAF
45.39	19.80	1.30	-3.48	-4.35(±0.2)	135Kot

* m NaF - concentration measured in quenched solutions.
145RAF-149RAF –P=170-180 MPa

In the region of low concentrations of NaF (<0.1 моль/кг H_2O) the dependence curve m Ta on m NaF may be more complex than it is given in equations 4 and 5. The data obtained are not sufficient for perfect description of solubility in diluted NaF solutions. The main purpose of these studies is to establish the boundaries of the fluid immiscibility in the system NaF-H_2O. In diluted solutions (m NaF < 0.1), as it was shown by thermodynamic calculations, the dependence of m Ta on m NaF is more complicated than equations 4 and 5, as HF, H_2O, and NaF participate in the reactions of microlite incongruent dissolution.

In the ranges from 1 to 11 wt% NaF at 170 MPa, from 5 to 25 wt% NaF at 200 MPa and from 15 to 25 wt% NaF at 230 MPa (Figure 2-a) an abnormal change of microlite solubility takes place: total (apparent) concentration of tantalum in quenched solution decreases with increasing total concentration of NaF. It is related to some critical phenomena in the system NaF-H_2O - with the transition of water-salt solution from the homogeneous field to the immiscibility field. As it was mentioned above (equation 3), dependences C_{Ta} on C_{NaF} in the immiscibility field of L_l and L_2 fluid phases are linear, whereas in the coordinates log m Ta – log m NaF (Figure 2-a) the dependence, of course, non-linear. The linear equations of experimental data of C_{Ta} depending on C_{NaF} in immiscibility field of L_l and L_2 fluid phases, for pressures of 170, 200, and 230 MPa take the forms:

$$C_{Ta} \times 10^5 \text{ (wt.\%)} = 5.782 - 0.223 \times C_{NaF} \text{ (wt.\%)} \pm 1.5 \quad (6)$$

$$C_{Ta} \times 10^5 \text{ (wt.\%)} = 10.494 - 0.295 \times C_{NaF} \text{ (wt.\%)} \pm 1.5 \quad (7)$$

$$C_{Ta} \times 10^4 \text{ (wt.\%)} = 13.459 - 0.457 \times C_{NaF} \text{ (wt.\%)} \pm 1.5 \quad (8)$$

The data obtained let to outline the immiscibility field in the system NaF-H_2O at 800°C. Compositions of the L_l and L_2 phases at pressures of 230, 200, and 170 MPa are 12 and 25 wt.%, 5 and 26 wt.%, 1 and 11 wt. % NaF, respectively. The phase diagram, presented in Figure 3, shows that the immiscibility field at 800°C is limited below by a pressure $p_{cr\ max} = 165 \pm 5$ MPa and above by a pressure $p_{cr\ min} = 236 \pm 5$ MPa. At these points L_l and L_2 phases have the same composition ($L_l = L_2$, 3 and 20 wt.% NaF).

Ca concentration in quenched solutions did not practically depend on NaF concentration and was $10^{-3\pm0.5}$ m Ca. Such a high concentration of Ca in the solution might have been related to the presence of an impurity in the distilled water because fluorite solubility in fluoride solutions is extremely low (Tropper 2007).

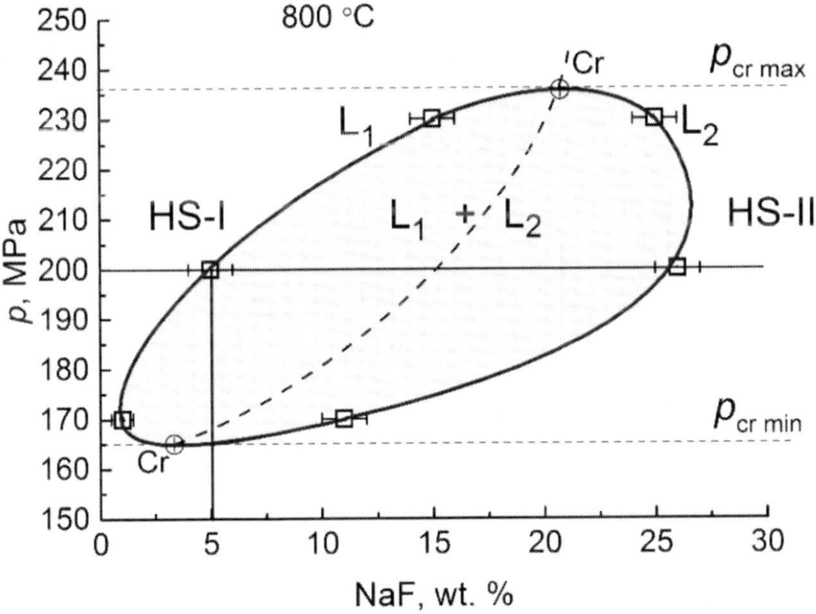

Figure 3. Immiscibility field ($L_1 + L_2$) in the system NaF–H_2O at 800°C on the experimental data. Symbols: HS-I is a homogeneous solution at low concentrations of NaF; HS-II is a homogeneous solution at high concentrations of NaF; $L_1 + L_2$ is the immiscibility field of fluid phases L_1 and L_2 in the system NaF–H_2O; $p_{cr\ min}$ and $p_{cr\ max}$ are min and max critical pressures on immiscibility field at the temperature.

NaF concentration in the fluid phase L_1 at 200 MPa was about 1.25 ± 0.26 mol·kg^{-1} H_2O (5 ± 1 wt.%), and the equilibrium with microlite and fluorite concentration of tantalum – 5.25 ×10^{-6} m Ta (log m Ta = –5.28). However, in a homogeneous solution (HS-I), containing 1.0 m NaF, tantalum concentration is 2.63× 10^{-6} mol·kg^{-1} H_2O. Assuming that at the phase boundary HS-I and L_1 solutions contain equal concentrations of NaF and tantalum, the HF concentration must be also equal. To determine m HF the experiments on microlite incongruent solubility were performed at 800°C and p = 200 MPa in 1.0 m NaF solution, containing varies concentrations of HF. The experimental results are shown in Table 2. Influence of m HF on tantalum content in 1.0 m NaF solution, equilibrium with microlite and fluorite, can be approximated (Figure 4-a) with the equation:

$$m\,Ta \times 10^5 = 0.27 + 4.359 \times m\,HF + 210.09 \times m\,HF^2 + 760.11 \times m\,HF^3 \quad (9)$$

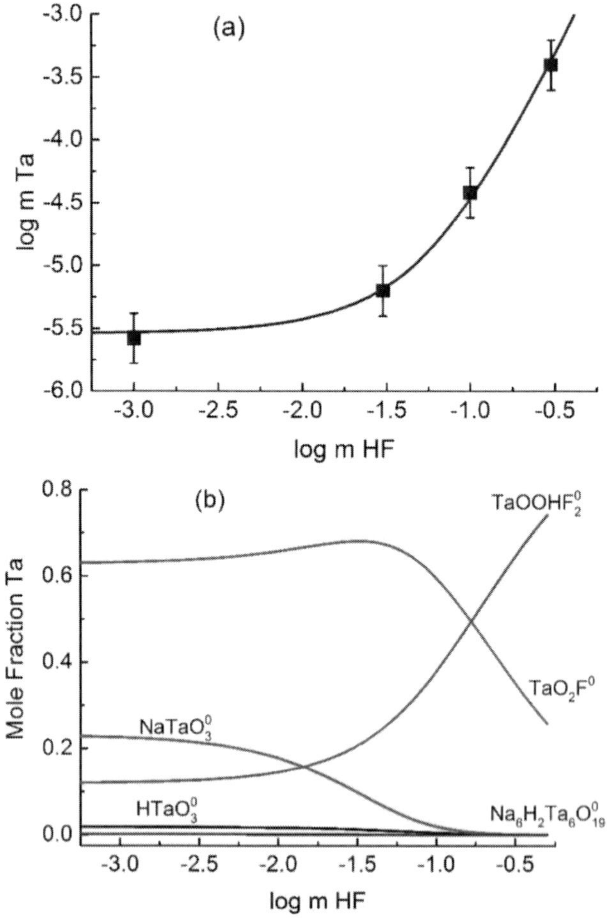

Figure 4. Effect of m HF on m Ta (a) and the distribution of Ta^{5+} species (b) in the system microlite-fluorite at 800°C, 200 MPa in solutions of 1.0 m NaF.

According to equation 7, the concentration of tantalum, equal to 5.25×10^{-6} m Ta, is achieved in 1.0 m NaF solutions containing 0.025 ± 0.003 m HF (log m HF = −1.62). Thus, the bundle of the homogeneous solution of sodium fluoride to two immiscible liquids leads to the enrichment of L_1 phase by HF and, respectively, NaOH accumulates in phase L_2.

Table 2. Effect of m NaF and HF on microlite solubility at 800°C and 200 MPa. Run duration 7-9 days

Initial solution				Quenched solution		Run
NaF wt.%	m NaF	HF (m)	log m NaF	log m Ca	log m Ta	no
0	0.014*	–	-1.858*		-6.31(±0.3)	1332Ta
0	0.011*	–	-1.940	-2.64	-6.18(±0.3)	143Ta
0	0.167*	–	-0.777*		-5.76(±0.2)	133Ta
0.418	0.100	–	-1.000		-5.54(±0.2)	128Ta
0.418	0.100	–	-1.000	-4.1	-6.08(±0.3)	136Ta
5.030	1.262	–	0.101		-5.64(±0.2)	124Ta
5.109	1.282	–	0.108		-5.22(±0.2)	134Ta
8.083	2.094	–	0.321		-5.39(±0.2)	130Ta
11.49	3.092	–	0.490		-5.38(±0.2)	125Ta
15.42	4.341	–	0.638		-5.42(±0.2)	126Ta
20.28	6.060	–	0.783		-5.46(±0.2)	127Ta
25.39	8.107	–	0.909		-5.66(±0.2)	129Ta
30.32	10.36	–	1.015		-5.40(±0.2)	131Ta
35.24	12.96	–	1.113		-4.30(±0.2)	132Ta
45.39	19.80	–	1.297		-4.35(±0.2)	135Ta
4.030	1.000	0.00	0.000	-3.43	-5.58(±0.2)	144Ta
4.030	1.000	0.03	0.000	-3.30	-5.20(±0.2)	140Ta
4.030	1.000	0.10	0.000	-3.16	-4.42(±0.2)	141Ta
4.030	1.000	0.30	0.000	-2.68	-3.40(±0.2)	142Ta

* m NaF - concentration measured in quenched solutions.

3.1.1. Ta^{5+} Speciation

We have done thermodynamic calculations for the homogeneous field of NaF solutions. Taking into account an inessential effect of pressure of homogeneous solution on microlite solubility in the range 170 – 230 MPa we combined the results of those investigations and all the calculations were done at 800°C and p = 200 MPa. Thermodynamic properties of sodium, calcium and fluorine species (Na^+, $NaOH°$, $NaF°$, Ca^{2+}, $CaOH^+$,

CaF^+, F^-, HF_2^-, $HF°$) were taken from the works (Johonson 1992, Shock 1997$_1$). In order to solve the inverse problem different possible complexes of Ta^{5+} both monomeric and polymeric ones were considered. A convergent solution has been only found for the following species of Ta^{5+}: $HTaO_3^0$, TaO_2F^0, $TaOOHF_2^0$, $NaTaO_3^0$, and $Na_6H_2Ta_6O_{19}^0$ (Figure 2-b). Experimental results at 800°C and p = 200 MPa in solutions containing 1.0 m NaF and different concentrations of HF have been used to refine the constant of formation of the specie $TaOOHF_2^0$. The model thermodynamic calculations indicate that the predominant Ta^{5+} species were $NaTaO_3^0$, TaO_2F^0, and $TaOOHF_2^0$ (Figure 4-b). The contribution of species $HTaO_3°$ was insignificant. As a result of the carried out thermodynamic calculations the constants of the following reactions have been estimated:

$$\text{Mic} + 2HF^0 = \text{Fl} + 2HTaO_3^0 + NaF^0 \quad (10)$$

$$\text{Mic} + 4HF^0 = \text{Fl} + 2TaO_2F^0 + NaF^0 + 2H_2O \quad (11)$$

$$\text{Mic} + 6HF^0 = \text{Fl} + 2TaOOHF_2^0 + NaF^0 + 2H_2O \quad (12)$$

$$\text{Mic} + NaF^0 = \text{Fl} + 2NaTaO_3^0 \quad (15)$$

$$3\text{Mic} + 3NaF^0 + H_2O = 3\text{Fl} + Na_6H_2Ta_6O_{19}^0 \quad (13)$$

It was found that negative logarithms of constants ($pK = -\log_{10} K$) of reactions 10–13 are equal to:

$$pK_{12} = 11.63,\ pK_{13} = 5.39,\ pK_{14} = 4.29,\ pK_{15} = 12.29,\ pK_{16} = 8.47.$$

The obtained reaction constants were used for model calculation of Mic solubility in NaF solutions (Figure 2-a). It was also found that equilibrium concentration of tantalum in the studied system has the minimum value $10^{-6.5\pm0.5}$ m Ta in 0.01 m NaF solution.

Table 3. Effect of m NaF on microlite solubility at 800°C and 230 MPa. Run duration 7–9 days

Initial solution			Quenched solution		Run
NaF wt.%	m NaF	log m NaF	log m Ca	log m Ta	no
0.00	0.016*	-1.80*		-6.95(±0.3)	117Ta
0.42	0.100	-1.00		-6.41(±0.3)	113Ta
2.06	0.50	-0.30		-6.05(±0.3)	116Ta
8.96	2.34	0.37		-5.15(±0.2)	118Ta
11.41	3.07	0.49		-5.42(±0.2)	119Ta
13.16	3.61	0.56		-4.35(±0.2)	104Ta
13.26	3.64	0.56		-5.16(±0.2)	114Ta
20.00	5.95	0.78		-4.49(±0.2)	115Ta
20.16	6.01	0.78		-4.48(±0.2)	105Ta
25.47	8.14	0.91		-5.02(±0.2)	106Ta
27.59	9.07	0.96	-3.07	-4.80(±0.2)	121Ta
33.89	12.21	1.09	-2.87	-4.74(±0.2)	122Ta
38.91	15.17	1.18	-3.13	-4.57(±0.2)	1232Ta
45.99	20.28	1.31	-2.78	-4.08(±0.2)	123Ta

* m NaF - concentration measured in quenched solutions.

3.2. Pyrochlore Solubility

We studied the equilibrium between pyrochlore and fluorite or the reaction of incongruent dissolution of pyrochlore with formation of fluorite which can be presented in generalized form by the following reaction:

$$Pcl + solution_{init} \leftrightarrow Fl + solution_{fin}(+species\ Nb^{5+}),$$

where $solution_{init}$ and $solution_{fin}$ –components of initial and final solution (HF, NaF, H_2O), which shoild be introduced in order to maintain the mass balance of all the elements. The investigations were carried out at 800°C and 200 MPa in the solutions containing 0.1 – 16.6 m NaF, and covered both homogeneous fields HS-I (0.1 -1.25 m NaF) and HS-II (m NaF > 7.9), and the fluid immiscibility field ($1.25 \leq$ m NaF ≤ 7.9). The analysis results of the solutions after the runs are given in Table 4. The data obtained (Redkin 2016_2) suggest the obvious decrease of m Nb on m NaF in the homogeneous field of the water-salt solution except for the field of fluid immiscibility L_1+L_2 (Figure 5), where apparent solubility of pyrochlore (C_{Nb}) has an inverse dependence on C_{NaF}.

$$C_{Nb} \times 10^3 \text{ (wt.\%)} = 15.128 - 0.572 \times C_{NaF} \text{ (wt.\%)} \pm 1.5 \quad (14)$$

3.2.1. Nb^{5+} Speciation

Using the software OptimA and experimental data on incongruent solubility of Pcl in homogeneous regions HS-I and HS-II, the inverse thermodynamic problem has been solved: the most predominant species of Nb^{5+} have been found and their stability constants have been calculated. The best convergence of calculated and experimental data was achieved when the following niobium species were used: $HNbO_3^0$, NbO_2F^0, $NbOOHF_2^0$ and $Na_6H_2Nb_6O_{19}^0$. The compositions of the species were chosen by analogy with the species of Ta^{5+} (Redkin $2015_{1,2}$, Redkin 2016). To identify the species $NbOOHF_2^0$ we used the results of experiments in solutions of 1.0 m NaF + HF (0.01-0.3 m) (Figure 6). Thermodynamic calculations show that NbO_2F^0 specie makes the main contribution to the gross concentration of Nb in the studied homogeneous solutions. The mole fraction of Nb in the form of $NbOOHF_2^0$ and $Na_6H_2Nb_6O_{19}^0$ species in NaF solutions containing m NaF > 10 does not exceed 20 mole %. As a result of the conducted thermodynamic calculations the constants of the following reactions have been calculated:

$$\text{Pcl} + 2HF^0 = \text{Fl} + 2HNbO_3^0 + NaF^0 \quad (15)$$

$$\text{Pcl} + 4HF^0 = \text{Fl} + 2NbO_2F^0 + NaF^0 + 2H_2O \quad (16)$$

$$\text{Pcl} + 6HF^0 = \text{Fl} + 2NbOOHF_2^0 + NaF^0 + 2H_2O \quad (17)$$

$3Pcl + 3NaF^0 + H_2O = 3Fl + Na_6H_2Nb_6O_{19}^0$ (18)

It was found that pK of reactions 15–18 are equal to:

$pK_{18} = 7.27$, $pK_{19} = 2.60$, $pK_{20} = 1.43$, $pK_{21} = 8.99$.

Table 4. Effect of m NaF on pyrochlore solubility at 800°C and 200 MPa. Run duration 7–9 days

Initial solution				Quenched solution		Run
NaF wt.%	m NaF	HF (m)	log m NaF	log m Ca	log m Nb (±0.2)	no
0.42	0.10		-1.00		-4.32	136Nb
0.83	0.20		-0.70		-5.18	137Nb
0.83	0.20		-0.70	-3.41	-3.96	148Nb
2.06	0.50		-0.30		-4.59	138Nb
3.00	0.74		-0.13		-4.23	139Nb
4.35	1.08		0.034	-2.84	-4.33	149Nb
4.35	1.08		0.034		4.93	150Nb
4.89	1.22		0.088		-4.19	141Nb
6.23	1.58		0.20	-2.87	-3.98	150Nb
7.43	1.91		0.28		-4.43	142Nb
12.27	3.33		0.52		-3.03	143Nb
20.19	6.03		0.78		-2.89	144Nb
25.46	8.13		0.91		-3.94	145Nb
30.33	10.37		1.01		-4.14	146Nb
40.23	16.03		1.21	-3.28	-3.49	147Nb
4.03	1.00	0.00	0.00		-4.40	140Nb
4.03	1.00	0.01	0.00	-2.35	-3.75	151Nb
4.03	1.00	0.03	0.00	-2.71	-3.54	152Nb
4.03	1.00	0.10	0.00		-3.45	153Nb
4.030	1.00	0.30	0.00	-2.23	-2.22	154Nb

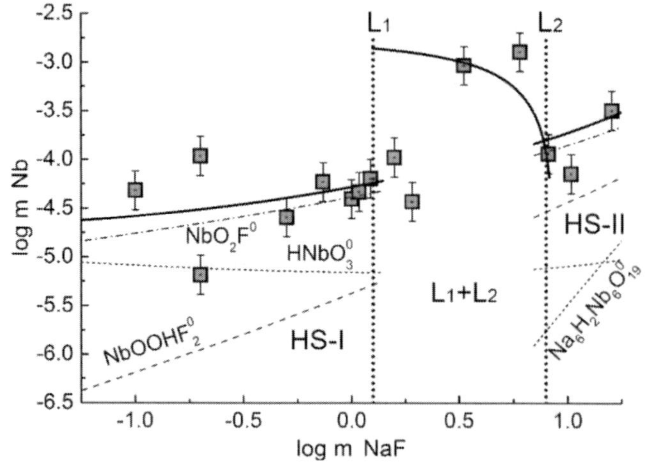

Figure 5. Effect of m NaF on m Nb and Nb^{5+} speciation in the system pyrochlore-fluorite at 800°C, 200 MPa in NaF solutions (experimental and calculated data). Solid line shows a change of log m Nb depending on log m NaF. Symbols: see Figure 3.

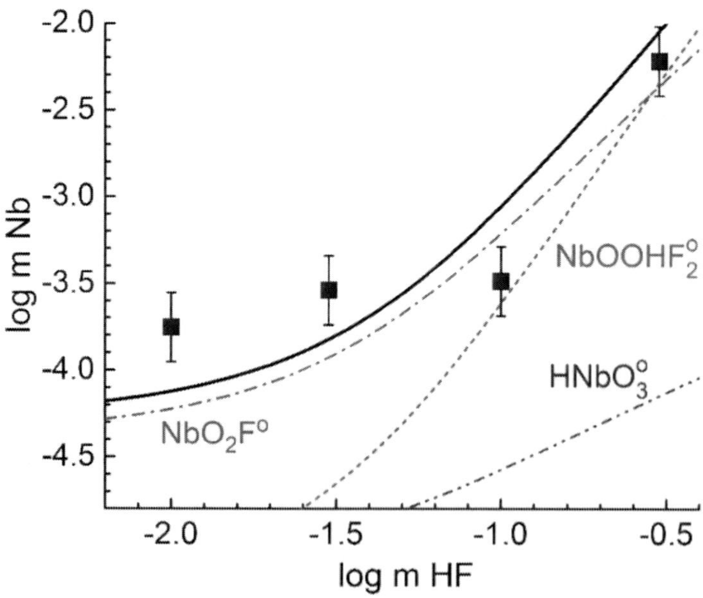

Figure 6. Effect of m HF on m Nb in the system pyrochlore-fluorite at 800°C, 200 MPa in solutions of 1.0 m NaF (experimental and calculated data). Solid line demonstrates a change log m Nb depending on log m NaF and the concentration of predominant Nb^{5+} species.

3.3. U-Pyrochlore and Uraninite Solubility

Uranium-pyrochlore (U-Pcl), $(Na_{1.2}Ca_{0.4}U_{0.3})Nb_2O_{6.4}F_{0.4}$, (Figure 7) was obtained by hydrothermal synthesis by substitution of Ca^{2+} for U^{4+} and Na^{+} in the cubic polyhedron of pyrochlore according to the reaction:

$$Pcl+0.3Urn+0.6NaF_{aq}+0.2H_2O \rightarrow U\text{-}Pcl+0.6Fl+0.4NaOH_{aq} \qquad (19)$$

The synthesis was performed at 800°C, 200 MPa, Co-CoO buffer in 1 m NaF solution saturated in uraninite (Urn). The composition of the obtained U-Pcl within the method error (electronic microscope = REM) correspotonded to the pyrochlore composition described earlier (Redkin 2013_1).

Figure 7. SEM images of the uranium-pyrochlore (U-Pcl) obtained at 800°C, 200 MPa. Both uraninite (Urn) and Fl are presented with synthetic U-Pcl.

Despite the presence of Urn and Fl (reaction 19, Figure 7, Figure 8) in U-Pcl, additional samples of Urn and Fl, in the amount of 10 and 1-2 mg accordingly were introduced in the experiments on solubility.

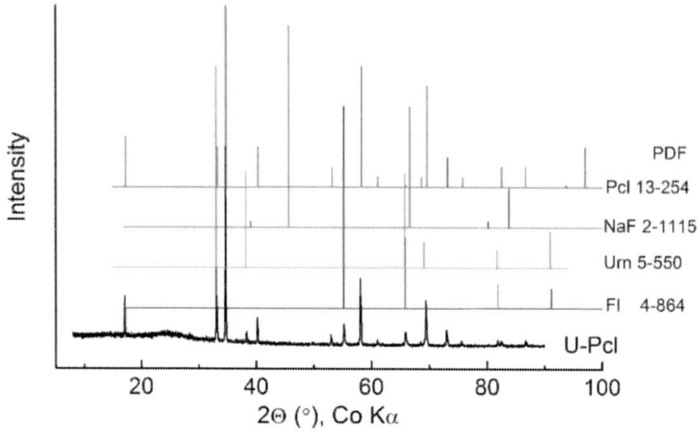

Figure 8. X-ray diffraction patterns of synthetic U-Pcl and PDF2 standards of Pcl, villiaumite (Vil), Urn, and Fl.

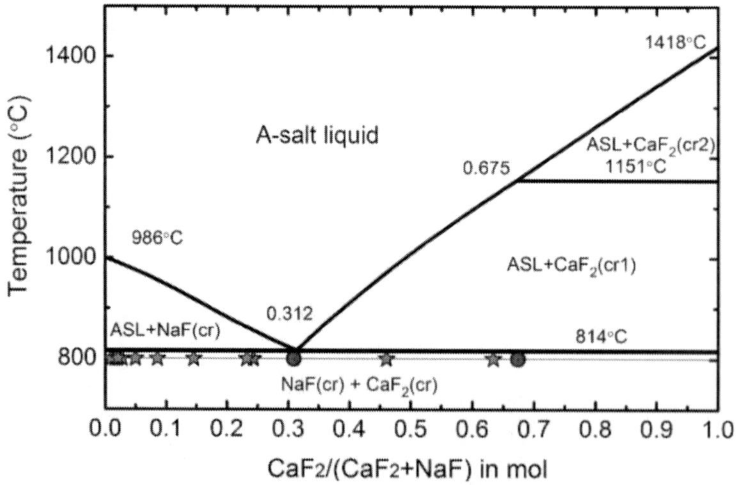

Figure 9. Variation of CaF_2 mol fraction in experiments at 800ºC, 200 MPa on the phase diagram in the system CaF_2-NaF from FTsalt-FACT salt database (Bale et al., 2016). Symbols: ☆ – experiments with UO_2, ⊙ - experiments in the system NaF-CaF_2-H_2O.

The ratio CaF_2/NaF varied within a wide range (shown in stars in Figure 9), so, the appearance of a melt containing both NaF and CaF_2 was expected. However, such glasses were not found in the run products. Melting at 800°C, 200 MPa was not observed either in check experiments (shown in circles in Figure 9), where the charge NaF – CaF_2 was used.

The studied solution of NaF, was simultaneously saturated in niobium and uranium. Uraninite controlled and at the same time saturated the solution with U^{4+} species and pyrochlore supplied Nb^{5+} species into the solution. Analysis results of the solutions for U and Nb content in the experiments on U-Pcl and Urn solubility in NaF solutions are presented in Table 5 and Figure 10. In the field of liquid immiscibility of the system NaF-H_2O two solutions (fluids) L_1 and L_2 of different density coexist simultaneously. Their ratio depends on gross of NaF. Saturration concentrations of the L_1 phase with uranium $C_U(L_1)$ and niobium $C_{Nb}(L_1)$ is higher than that $C_U(L_2)$ $C_{Nb}(L_2)$ of the L_2 phase. So apparent solubility of Urn and U-Pcl in the field $L_1 + L_2$ has an inverse dependence on NaF concentration. It should be noted that linear character of the dependence of apparent solubility (concentration) of niobium and uraniun on mass concentration of NaF in the system containing uranium is disrupted because in the field of fluid immiscibility of L_1+L_2 a double salt of sodium and uranium fluorides (NUF) is observed in the quenched products of the runs. The interaction of uraninite with sodium fluoride caused a notable change of the ratio NaF/NaOH in the solution which could lead to the change of the fluid immiscibility boundaries. Actually, instead of the double system NaF-H_2O we obtained the triple system NaF-NaOH-H_2O.

Diffraction pattern of solid runs products concerning the interaction of Urn with acid solution of 1m NaF + 0.3 m HF (#3NUF) is given in Figure 11. According to XRD phase analysis significant amounts of uraninite were noted in experiment products (comparison pattern PDF 4+: 01-071-6416) (5-12%) as well as monohydrate of double salt of uranium tetrafluoride and sodium fluoride (sodium penta-fluoro-uranate (IV) hydrate = Sodium Uranium Fluoride Hydrate) $Na[UF_5]\cdot H_2O$ (PDF 4+: 01-073-9350) (88-95%).

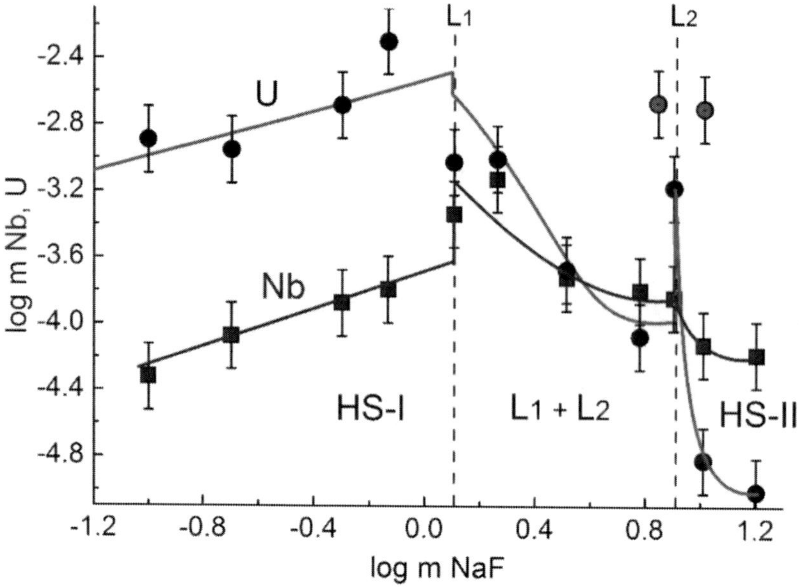

Figure 10. Effect of m NaF on m Nb and m U in solution equilibrated with U-Pcl, Urn, and Fl at 800°C, 200 MPa.

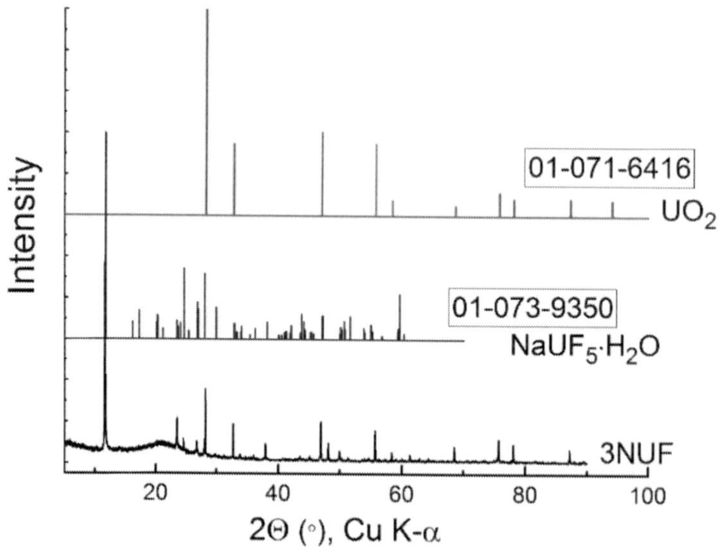

Figure 11. X-ray diffraction patterns of solids obtained in experiment #3NUF on Urn solubility in 1.0 m NaF + 0.3 m HF solution at 800°C, 200 MPa.

The latter compound according to (Fisher 2004), crystallizes in the orthorhombic lattice space group Pbcn, and unit cell parameters are a = 6.9928, b = 8.7617, c = 15.10 Å. Parameter "*c*" was slightly corrected to have a better agreement of calculated interplane distances "*d*" with the XRD data (Figure 11). Water free Na[UF_5] (reference standard PDF 4+: 00-012-0047) is not identified in the XRD pattern (Figure 11).

Table 5. Effect of m NaF on U-pyrochlore and uraninite solubility at 800°C and 200 MPa, and the CCO buffer. Run duration 7–9 days

Initial solution		**Quenched solution**			**Run**
m NaF	**HF (m)**	**log m Ca**	**log m Nb**	**log m U**	**#**
0.10	0.00	-3.05	-4.32	-2.89	1UN
0.20	0.00	-3.35	-4.08	-2.95	2UN
3.26	0.00	-2.85	-3.72	-3.67	3UN
10.23	0.00	-3.10	-4.12	-4.82	4UN
0.50	0.00	-3.29	-3.88	-2.68	5UN
1.83	0.00	-2.60	-3.13	-3.01	6UN
1.27	0.00	-2.66	-3.34	-3.02	7UN
6.04	0.00	-2.81	-3.80	-4.07	8UN
0.74	0.00	-3.07	-3.79	-2.30	9UN
16.01	0.00	-3.03	-4.19	-5.01	10UN
8.00	0.00	-2.88	-3.84	-3.18	11UN
10.36	0.00	-	-	-2.70	1Urn+Qtz
7.00	0.00	-	-	-2.67	2Urn+Qtz
1.00	0.30	-	-	-2.50	3NUF
1.00	0.10	-	-	-3.58	4NUF

In the run products washed out of the platinum capsule alongside with black Urn crystals there were noted newly formed emerald-green plate-shaped crystals. Emerald-green crystals become light under water evaporation from Petri dish in desiccator at 100°C. Corroded Urn crystals and newly formed aggregates of NUF crystals are distinctly seen in the picture of run products (Figure 12) taken with electronic microscope. It was performed the photography of crystals stuck on the conducting rubber with gold spraying 3-5 nm thick. Surface compositions of the indicated uranium phases are shown on the triangular diagram (Figure 13) in Na-U-F coordinates. The data obtained show that uraninite surface (light crystals) is subjected to solubility and formation of fluoride compounds takes place on it. However, the XRD pattern Figure 11 does not discover essential amounts (3-5%) of other fluorides of uranium and their hydrates. According to SEM analysis the surface layer of NUF is deficient in Na compared to an ideal formula $Na[UF_5]\cdot H_2O$. The lowest Na concentrations are noted in the samples made in a point, whereas the scanning of the surfaces allowed finding higher sodium concentrations. Obviously, in the course of analysis of the NUF crystal surface the excitation band is heated and sodium ions and (or) NaF molecules migrate from the hot zone into the cold one. Thus all the NUF compositions are inside the triangle UF_4-$NaUF_5$-UOF_2 (Figure 14).

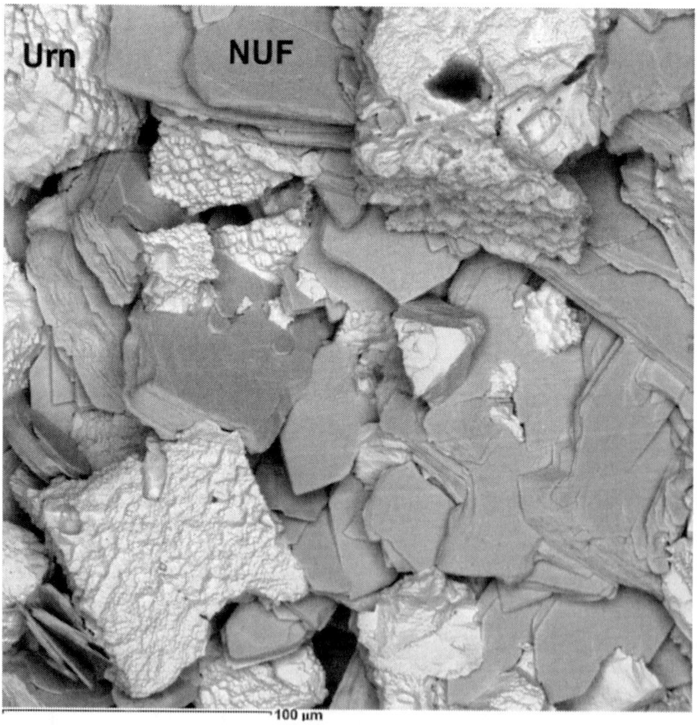

Figure 12. SEM images of Urn and its derivates (NUF) obtained in experiment #3NUF on Urn solubility in 1.0 m NaF + 0.3 m HF solution at 800°C, 200 MPa.

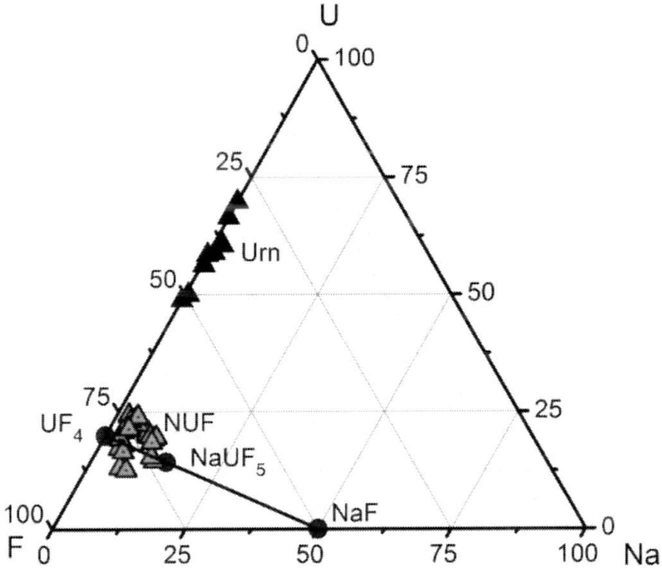

Figure 13. Triangular diagram of the compositions of Urn (black symbols) and NUF (green triangles) in coordinates Na-U-F by SEM analysis after run at 800°C, 200 MPa in solution of 1.0 m NaF + 0.3 m HF.

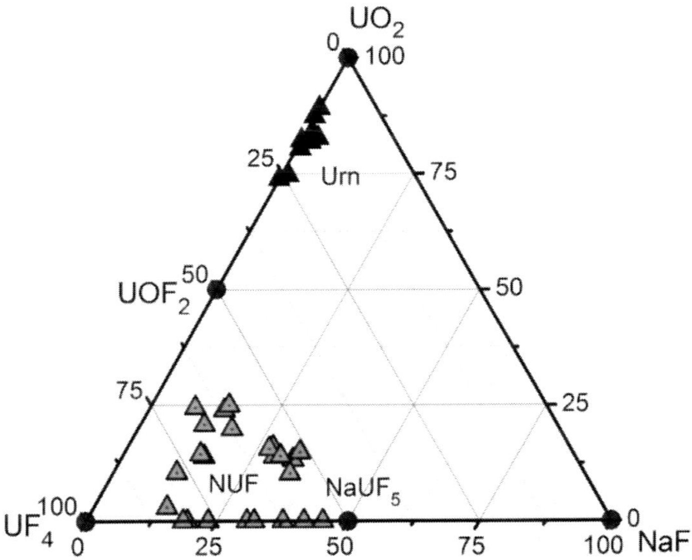

Figure 14. Triangular diagram of the compositions of Urn (black symbols) and NUF (green triangles) in NaF-UO_2-UF_4 coordinates (recalculated after Figure 13).

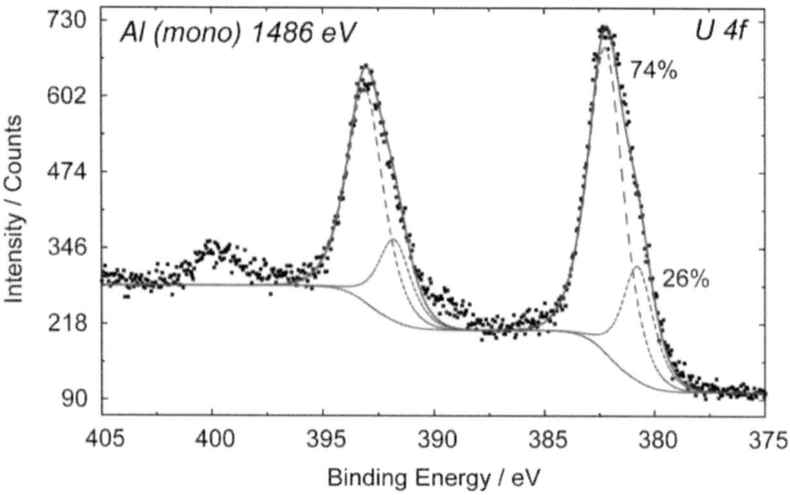

Figure 15. U4f XPS surface scan of the sample # 3NUF; curve fitting results.

According to the XPS analysis the run products are mainly compounds of U^{4+} with binding energies (BE) U4f7/2 at 381.1 and 382.5 eV, and U4f5/2 at 392.0 and 393.5 eV (Figure 15), corresponding to U^{4+} state in such compounds as UO_2 and UF_4, (including $NaUF_5$), respectively (Schindler 2009, Ilton 2011, Yeon 2015). The excitation band of the device is wide and captures both uraninite and newly formed NAU phase.

The compound of $NaUF_5$ composition has be mentions in literature several times. In (Martin 1951) it is noted that double salt of uranium tetrafluoride and fluoride of green color is formed at t > 450°C as a by-product at decomposition of a complex uranium salt ($3NaF{\cdot}UF_6$) according to the scheme:

$$3NaF{\cdot}UF_6 \rightarrow NaUF_6 + 2\ NaF + \tfrac{1}{2}\ F_2 \rightarrow NaUF_5 + 2\ NaF + F_2 \qquad (20)$$

In the review (Grenth 2010) the compound $NaUF_5$ is mentioned as an actually existent one but there are no available data on it. We also noted (Redkin 2013_2) formation of uranium fluorides in hydrothermal systems containing aluminosilicate melts and high concentrations of sodium, potassium, lithium and aluminum fluorides. According to the SEM analysis NUF composition corresponds to the formula $NaUF_{5+x}$.

The analysis was carried out on polished surfaces with carbon deposition of 15-20 nm thick. As a standard for fluorine, natural sellaite MgF_2 and a monocrystalline villiomite grown from NaF melt were used.

Binary compound $NaF{\cdot}UF_4$ at 800°C should be represented by a melt (Thoma 1957, Barton 1958). If assumed that in the capsules containing 10 mg UO_2 and 0.15 ml solution 1.0 m NaF + 0.3 m HF, in the experiment UF_4 was formed and in the system $NaF\text{-}UF_4\text{-}H_2O$ melting took place, melt composition could vary greatly. It is unlikely that during 3 minutes (just such time is necessary for the reactor with the run capsule cooling up to the room temperature) the crystallization of melt with formation fixed composition NUF, water-free $NaUF_5$ or crystal-hydrate $Na[UF_5]{\cdot}H_2O$ took place. Thus it could be supposed that on the surface of uraninite $UO_{2.0}$ there was a melt in which mole ratio NaF/UF_4 was within the range from 7/6 to 1/2.

3.3.1. U^{4+} speciation

In order to determine the composition of uranium complexes we only used the data of the experiments in which formation of NUF phases did not take place. The experiments with UO_2 conducted in acid solutions (1.0 m NaF + 0.1 m HF (#4NUF) and 1.0 m NaF+0.3 m HF (#3NUF)) and in highly concentrated NaF solutions containing SiO_2 additive were not considered either because significant amounts of NUF were detected in the run products. Using the software OptimA and experimental data on UO_2 solubility in the homogeneous field HS-I it was established that the major Ue^{4+}species in the solutions containing 0.1 – 1.0 m NaF is fluoride complex $UOOHF^0$ (or $U(OH)_3F^0$). According to calculations the equilibrium constant of reaction (21)

$$UO_2cr + HF^0aq = UOOHF^0aq \qquad (21)$$

at 800°C and 200 MPa is Kp(24) = $mUOOHF^0aq/mHF^0aq$ = 0.11. The calculation results are presented in Figure 9.

4. Discussion

Experimental data and calculated based on them constants of formation of hydroxo- and fluoride complexes of tantalum, niobium and uranium at 800°C and 170-230 MPa are important for the study of geochemistry of these elements in magmatic processes. Fluoride solutions are considered as the most probable medium for the hydrothermal transfer of tantalum, niobium and uranium. T-P parameters chosen for the experiments cover both the region of homogeneous solutions and a wide range of immiscibility in the system NaF-H_2O.

4.1. Thermodynamic Properties of the Species Nb^{5+}, Ta^{5+} и U^{4+}

After the analysis of the state of affairs with rare metals (Wood 2004) a number of works were published on solubility of the niobium oxides (Peiffert 2010, Akinfiev 2015, Timofeev 2015, Lukyanova 2017) and the tantalum oxides (Zaraisky 2010, Timofeev 2017). Based on those studies thermodynamic properties of hydroxo- and hydroxo- fluoride complexes have been calculated.

Studying B-Nb_2O_5 solubility in perchlorite solutions at 10-70°C in a wide range pH and ionic strength Peiffert with collaborators (Peiffert 2010) concluded that in case of noncomplexation with ClO_4^- the predominant Nb^{5+} complexes should be hydroxo complexes $Nb(OH)_n^{5-n}$ (were $n = 4$-7). Even in acid solutions (pH = 1.0), niobium concentration equilibrium with Nb_2O_5 was extremely low and polymeric complexes (Etxebarria 1994) $[H_{3-n}Nb_6O_{19}]^{(8-n)-}$ ($n = 0$-3) were practically absent.

In the work (Timofeev 2015) Nb_2O_5 solubility was studied in the solutions HF-$HClO_4$ at 150, 200, and 250°C and saturated water pressure. Analyzing the data obtained the authors calculated formation constants of two prevailing complexes of Nb^{5+}: $Nb(OH)^{4+}$ and $NbF_2(OH)_3^0$.

Of major interest are the data (Akinfiev 2015, Lukyanova 2017). Using experimental data on Nb_2O_5 solubility at 25°C, 1 bar (Peiffert 2010), 150-250°C and saturated water pressure (Timofeev 2015), 300-550°C and a pressure of 100 MPa in different solutions (Zaraisky 2010), a quantum-chemical approach to describe the properties of isolated molecules and software OptimA and OptimB [Shvarov, 2015], standard thermodynamic properties of niobium hydroxo- and hydroxo-fluoride complexes, HKL parameters of species (Johnson 1992) have been calculated. As pyrochlore solubility at 800°C, 200 MPa is known (Figure 5), and the chemical system is determined, then using those thermodynamic data the constant of reaction (18) and free energy of Pcl have been calculated. Although thermodynamic properties of species were outside the recommended range of T-P parameters the use of the obtained system of agreed data is efficient for calculations in the range HS-I from 0.1 to 1.0 m NaF. The predominant niobium species in water fluoride solution is $Nb(OH)_4F^0$. Mole fraction of the second significant species $Nb(OH)_5F^-$ does not exceed 0.1.

The investigation of the effect of m HF and ionic strength on $Ta_2O_5^{cryst}$ solubility was carried out in the work (Timofeev 2017) at 100-250°C and saturated water pressure. The authors showed that equilibrium between the solution of 0.075 m HF and solid Ta_2O_5phase occurs in 6 days at 100°C. Hydroxo-complex $Ta(OH)_5^0$ predominates in acid solutions (pH~1-3), containing low concentrations of HF. The species $Ta(OH)_3F_3^-$ predominates in solutions containing significant concentrations of HF. The authors suggest that at low temperatures (100-150°C) and pH, TaF_5^0 is formed together with the mentioned tantalum species. Equilibrium constants between Ta_2O_5 and the indicated species have been calculated.

Using experimental data (Zaraisky 2010) on Ta_2O_5 solubility in the temperature range 300-550°C and pressure of 100 MPa, thermodynamic properties of hydroxo- and hydroxo-fluoride species of tantalum have been calculated (Akinfiev et al., 2015). As in the case of pyrochlore we determined Gibbs free energy of microlite formation at 800°C and 200 MPa. Leveled calculation values of Mic solubility (Figure 2-a) and free energies of tantalum species at 800°C and 200 MPa were assumed as a basis. As a result, it was determined that the species $Ta(OH)_4F^0$ predominates at low m NaF (0.001-0.2 m). An increase of m NaF (to 1.0 m) leads to an increase of ionic strength and growth of $Ta(OH)_5F^-$ species part. It should be noted that using all four tantalum species ($Ta(OH)_5^0$,

$Ta(OH)_4F^0$, $Ta(OH)_3F_2^0$, and $Ta(OH)_5F^-$) in thermodynamic calculations does not give a good agreement with calculated data presented in Figure 2-a. The best agreement was observed when $Ta(OH)_5F^-$ species was excluded from model calculations. It is known that temperature increase leads to the growth of the fraction of uncharged species.

Thus, the progress in the study of thermodynamic properties of tantalum and niobium species is obvious for the conditions covering the field of hydrothermal ore formation (≤ 600°C). Unfortunately, at present there are no available data on the Gibbs free energies of formation of such ore minerals as pyrochlore, microlite, columbite, and tantalite. Therefore, using those thermodynamic data for the description of natural processes is still unobvious. Constants of reactions 12-16 and 18-21 can be recommended for model calculations at 800°C.

Although uranium is a strategic metal our knowledge about fluoride complexes of uranium (IV) particularly at high temperatures and pressures are quite fragmentary. Presented in the review (Grenth 1992) thermodynamic properties for uranium water species as a rule refer to standard conditions. The reliable enough thermodynamic data (HKF parameters) are available for the hydroxocomplex $U(OH)_4^0$, which prevails in diluted solutions (Shock 1997_2, Redkin 2013_2). The data are used for the calculations of UO_2 solubility in water and agree with the experimental data (Parks 1988, Redkin 1989, Redkin 1996) for high temperatures and pressures. The concentration of $U(OH)_4^0$ (or $UO_2(aq)$) in the temperature range 25-600°C, and a pressure of P_{sat} - 100 MPa does not exceed 10^{-8} $mol \cdot kg^{-1}$ H_2O.

An important part of halides in the mobilization of U(IV) into a hydrothermal solution is considered in (Zharikov 1987). The study was carried out with solutions containing relatively high concentrations of acids (HCl, HF) and salts (NaCl, KCl), in which solubility of uranium oxides was higher than sensitivity ($2 \cdot 10^{-6}$ m U_{total}) of spectroscopic Arsenazo-III method. The results obtained made it possible to show the effect of halide concentrations on UO_2 solubility in solutions and estimate the constants of uranium distribution between solutions and melts at different redox conditions.

Keppler & Wylle (Keppler 1990, 1991), studied uranium distribution between haplogranite melt and HF solution at 750°C, 100 MPa and Ni-NiO buffer and determined the ratio F:U = 2:1 in uranium complexes and concluded that UO_2F_2 is a prevailing species i.e., uranium is present in the solution even under reducing conditions as U(VI). Later (Peifert 1996), studied uranium solubility in halide solutions and haplogranite melt suggested that uranium should form two types of strong fluoride complexes with образует U(IV) and U(VI) in reducing and oxidizing conditions:

$$U^{4+} + mF^{-} = UF_m^{4-m} \ (m = 1\text{-}5) \quad (25)$$

$$UO_2^{2+} + nF = UO_2F_n^{2-n} \ (n = 1\text{-}4) \quad (26)$$

A significant contribution to the geochemistry of uranium in fluoride (HF) solutions was done by (Kovalenko 2012). They were the first to estimate equilibrium constants of the reactions

$$UO_2 + nHF + (2\text{-}n)H_2O = U(OH)_{4\text{-}n}F_n \ (n = 1\text{-}4) \quad (27)$$

and free energies of formation of uranium hydroxo-fluoride complexes (IV) at 500°C, 100 MPa. Bases on those data and the data obtained before by (Kovalenko 2011) it was shown that species U(IV) with fluorine are much stable than with and $U(OH)_3F^0$ species prevail in solutions containing up to 0.5 m HF and saturated with uraninite.

The analysis performed shows that the data on the forms of uranium transfer under high temperature conditions (>350°C) are rather scarce. It is the main reason why we do not consider magmatogenic uranium source significant for the formation of large and gigantic uranium deposits. Low temperature region (to 150°C), on the contrary is well studied. Suggested earlier (Naumov 1959, 1978) model of uranium transfer as carbonate uranile complexes is widely used for the construction of formation models of *unconformity-related uranium deposits* Athabasca (Canada), Jabiluka (Australia), Falea (Mali), Karhu (Russia) and so on (IAEA 2009). At the same time many hydrothermal uranium deposits are confined to beresites, albitites and other hydrothermal changes might form at 200-450°C. The available thermodynamic data are no enough to construct qualitative models for these deposits and estimation of ultimate uranium concentrations in ore-bearing solutions is made based on the results of experimental investigations.

4.2. Feature of the Immiscibility Field

Study results given in section 3 show the peculiarity of fluids L_1 and L_2 in the immiscibility field. Decay of the homogeneous solution of sodium fluoride into two immiscible liquids leads to the enrichment of phase L_1 with HF and, accordingly, to the accumulation of NaOH in L_2 phase. Three possible mechanisms for the accumulation of HF in the L_1 fluid phase were considered:

First, it was suggested that as a result of the NaF hydrolysis reaction

$$NaF + H_2O = HF + NaOH \quad (28)$$

An easily volatile HF component is formed which is redistributed into the low-density L_1 phase, while the dense L_2 fluid is enriched with alkali-NaOH. Accordingly (Veksler 2004), high temperature salt (including NaF) hydrolysis reactions enable physical separation of the hydrolysate by phases: HF goes in "vapor" (L_1), as well as NaOH redistributes in "melt" (L_2). In the immiscibility field L_1 and L_2 fluid phases, formed by the decay of NaF homogeneous solution, have a constant composition, and changes only the ratio of these phases, depending on the total concentration of NaF. In the case of HF redistribution between liquid phases of different density, as it approaches the border HS-I, the fraction of phase L_1 increases and L_2 phase fraction decreases. As a result, this leads to an increase in NaOH concentration in L_2 phase, so that the latter "drops" of L_2 phase will consist of a concentrated solution of NaOH. In this case Mic dissolution in L_1 phase leads to Ta_2O_5 formation, whereas sodium tantalates ($NaTaO_3$,

Na_2TaO_4 ...) will be formed in L_2 phase. Away from the boundary of HS-I, the L_2 phase fraction increases, in which hydrolysis of NaF would increase the content of HF in the L_1 phase and as a consequence, increase of Mic solubility. In fact, in immiscibility field the Mic apparent (total) solubility reduces with the growth of the total concentration of NaF. This concept is not consistent.

The second mechanism is the formation of HF in the interaction of Mic with NaF. This mechanism is also not well-founded, because Mic solubility in L_1 phase does not exceed 10^{-5} mol·kg^{-1} and an amount of dissolved Mic is not capable of producing 0.025 m HF. Furthermore, the reactions of the Mic incongruent dissolution in an aqueous solution of NaF forming fluorite and the prevailing species Ta^{5+} ($HTaO_3^0$, TaO_2F^0, $TaOOHF_2^0$, $NaTaO_3^0$ and $Na_6H_2Ta_6O_{19}^0$) or do not require HF, or lead to reduction of HF amount.

The third mechanism is related to the NaF hydrolysis features in the L_1 and L_2 phases. The properties of salt solutions much differ from those of gas or liquid. First of all it relate to properties of water, hydrogen bonding (Gorbaty 1995, Kalinichev 1994 Kalinichev 2017), ions and molecules hydration. Clasters $(H_2O)_n$ distribution in supercritical water (Figure 16) and in coexisting L_1 and L_2 fluids at 800°C and 200 MPa may be noticeably different. Thus, the fluids from immiscibility field have special properties. In the phase L_1, which exhibits the properties of the vapor phase together with the reaction 28, there is a reaction (29):

$$2NaF^0_{L1} + H_2O_{L1} = 2HF^0_{L1} + Na_2O^0_{L1} \quad (29)$$

leading to the formation of excess of the HF acidic component, relative to Na_2O alkali component. The reaction (24), leading to the formation of alkali component excess, also takes place in the dense (brine) L_2 phase, in addition to the reaction 30:

$$2NaF^0_{L2} + H_2O_{L2} = H_2F_2{}^0_{L2} + 2NaOH^0_{L2} \quad (30)$$

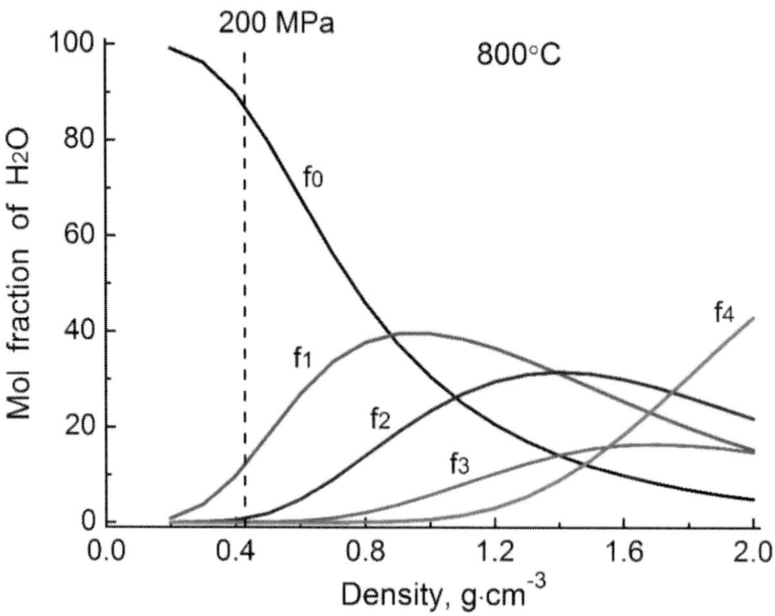

Figure 16. Effect of water density on fractions H_2O involved in clasters at 800ºC. Color lines indicate hydrogen bond fractions f_i with i = 0, 1, 2, 3, and 4, respectively. Calculated after (Kalinichev & Churakov, 1999).

As a result, L_l phase exhibits acidic properties, as L_2 phase is characterized by an excess of alkali (NaOH). Such a mechanism of accumulation of excessive amounts of 0.025 m HF in the L_l phase is the most likely.

4.3. Applications to Magmatic Systems

Juvenile granite magmas as a rule contain low clarkes of rare and scattered elements but can be a source of so-called «juvenile» hydrothermal solutions. Those solutions interact with high differentiated melts from which rare-metal Li-F granites form may lead to the loss of some ore elements and correspondingly increase the content of the others. An important task is the determination of the ratio of linear sizes (diameter) of the source of magmatogenic fluid and ore-enclosing rare-metal granite (or ore-bearing magmatic melt) which makes it possible a complete loss of U and Nb.

The following assumptions are taken as boundary conditions:

1. The initial melt contains tantalum, niobium, and uranium 100 g/т (=100 ppm) each in the form of microlite, pyrochlore and uraninite. i.e., it is a highly-differentiated melt oversaturated with tantalum, niobium and uranium.
2. Water content in granite melts at 800°C, 200 MPa reaches 8 wt%, i.e., a ton of the melt contains 80 kg H_2O or its solution and only 1/10 of the solution can participate in the processes of hydrothermal interaction with highly-differentiated melts and ore minerals.
3. The affecting solution is 0.1 m NaF.

The first two assumptions are out of doubt the third one is however disputable. Actually, solutions equilibrium with aluminosilicate melts such as granite magmas are relatively diluted by water solutions but contain a number of ore-forming elements: Na, K, Ca, Si, Al, Cl, F and others. However, the major components are sodium salts. Acidity of such solutions which is determined from the content of HCl и HF, NaOH in them is close the neutral at 800°C, 200 MPa. It should be noted that dissolving ore minerals are Na, Ca, Ta, Nb, U, F, and the effect of other elements on the solubility of Mic, Pcl, Urn in inessential. So it is assumed as a first approximation that hydtothermal fluid of magmatic origin only contains NaF.

Thermodynamic calculations were done using utility of Main HCh program (Shvarov 2008). Basic species were borrowed from the data bases Unitherm. The consistent values of Gibbs free energies for the species of niobium, tantalum, uranium (IV) and solid phases of pyrochlore and microlite were taken from this work (calculated from the final constants of reactions 12-16, 18-21, 24). Their extremes were used as solid phases although it is known that pyrochlore and microlite for a continuous series of solid solutions, they also can contain significant (to 0.3 mole/per formula of Pcl and Mic) amounts of uranium (Redkin 2010, Redkin 2013). We did not introduce solid solutions in our calculations so as not to complicate the task because for the case of ideal isomorphic miscibility of Nb and Ta, calculation results for m_{Nb} and m_{Ta} will change insignificantly. Although uranium entering into pyrochlores is able to influence pyrochlore solubility (see Figure 5 and 10), m Nb content in the solution equilibrium with U-Pcl remains 1.5 orders of magnitude lower than that of m U. Obviously uranium will be removed from pyrochlore sooner than niobium. Using the method of consequent reactors (changing the ratio sample/solution) the ratios of Mic, Pcl, Urn have been determined in the equilibrium system after the interaction of the above ore minerals with 0.1 m NaF solution. The calculation results were transformed to the volumes of melts of solution source and ore-bearing melt and then to their diameters (Figure 17).

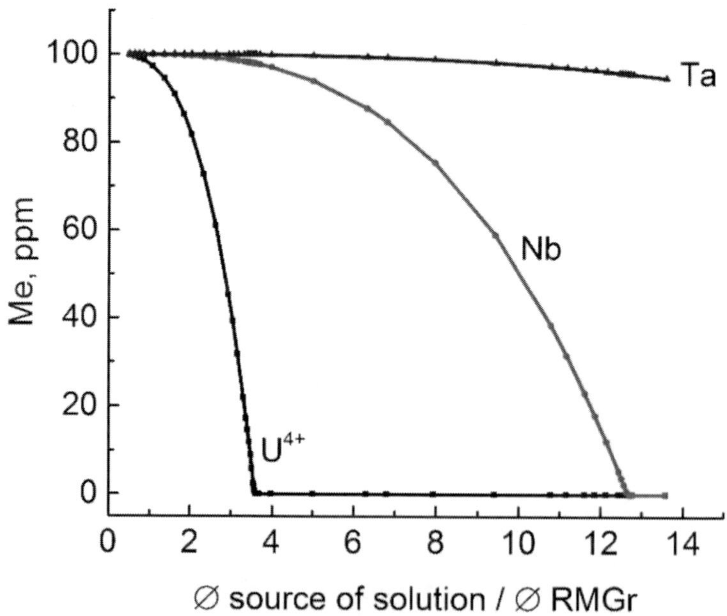

Figure 17. Effect of linear dimensions (Ø is diameter) of magmatic solution source and rare-metal granite (RMGr) on Ta^{5+}, Nb^{5+}, and U^{4+} removal from Mic, Pcl, and Urn bearing melt at 800°C, 200 MPa in 0.1 m NaF.

The calculations show that uranium the first of ore elements to be lost, then uranium and the last niobium. 3.6 -fold diameter of solution source and 46-fold volume is necessary for the complete loss of uranium under the conditions when only U^{4+}complexes are stable. To remove niobium from rare-metal ores initially containing niobium and tantalum 100 ppm each much more significant volumes of magmatogenic solutions are needed. Therefore, an essential amount of niobium is always present even in so-called tantalum ores.

CONCLUSION

The question of the magmatic source of ore matter for some deposits of rare and scattered elements is controversial. The main obstacle to the development of this scientific problem is the absence or lack of data on complex formation and limiting solubility of ore elements in magmatic hydrothermal solutions. To form large (industrial deposits) of rare and scattered elements, as a rule, acidic or alkaline solutions of high concentration are required). Such unusual solutions can be formed only in the immiscibility field of the water-salt solution at high T-P parameters. In this regard, our experimental studies of the solubility of ore minerals in the field of fluid immiscibility are of great interest for solving the problem of the genetic relationship between mineralization and magmatism and the construction of quantitative models for the formation of ore deposits.

ACKNOWLEDGMENTS

The authors are grateful to Karandashev V.K. (IMT RAS), Nekrasov A.N., Samokhvalova O.L., Dokina T.N. (IEM RAS) for the help in the study. We thank Ms. T. Kazachenko for editorial assistance. The work was supported by the Russian Foundation for Basic Research project 15-05-03393 and Program ONZ RAN.

REFERENCES

[1] Akinfiev, N. N., Zotov, A. V., Kotova, N. P., Korzhinskaya, V. S., & Lukyanova, E. V. (2015). "Nb and Ta in hydrothermal systems: thermodynamic description of hydroxo and hydroxofluoride complexes in wide range of temperature and pressure". Materials of a conference: Deposits of strategic metals: patterns of location, sources of matter, conditions and mechanisms of formation. *All-Russian conference dedicated 85th Anniversary of IGEM RAS foundation*. Moscow: IGEM RAS., P. 107-108 (in Russian). (Akinfiev 2015).

[2] Aksyuk, Anatoly M. (2009). *Behavior of fluorine in deep hydrothermal fluids and surface waters (Experimental study).* Summary of doctoral dissertation. Moscow., 59 p. (in Russian). (Aksyuk 2009).

[3] Bale, C. W., Bélisle, E., Chartrand, P., Decterov, S. A., Eriksson, G., Gheribi, A. E., Hack, K., Jung, I. H., Kang, Y. B., Melançon, J., Pelton, A. D., Petersen, S., Robelin, C., Sangster, J., Spencer, P. & Van Ende, M. A. (2016). Reprint of: FactSage thermochemical software and databases, 2010–2016. *Calphad.*, *55*, 1-19. 1. https://doi.org/10.1016/j.calphad.2016.05.002 December 2016. FTsalt - FACT Available from <http:// www.crct.polymtl.ca/ fact/ phase_ diagram.php?file=CaF2-NaF.jpg&dir=FTsalt> (Bale 2016).

[4] Barton, C. J., Friedman, H. A., Grimes, W. R., Insley, H., Moore, R. E. & Thoma, R. E. (1958). Phase Equilibria in the Alkali Fluoride – Uranium Tetrafluoride Fused Salt Systems: 1. The Systems $LiF–UF_4$ and $NaF–UF_4$. *J. Am. Ceram. Soc.*, *41*, 63–69. DOI: 10.1111/j.1151-2916.1958.tb13520.x Accessed on line 2 June 2006 (Barton 1958).

[5] Dolejš, D. & Baker, D. R. (2007). Liquidus equilibria in the system K_2O-Na_2O-Al_2O_3-SiO_2-F_2O_{-1}-H_2O to100MPa: I. Silicate-fluoride liquid immiscibility in anhydrous systems. *J. Petrol.*, *48*, 785-806. Accessed on line February 22, 2007 doi:10.1093/petrology/egm001 (Dolejš 2007).

[6] Etxebarria, Nestor E., Fernández, Luis A. & Madariaga, Juan M. (1994). On the hydrolysis of niobium(V) and tantalum(V) in 3

mol·dm−3 KCl at 25°C. Part 1. Construction of a thermodynamic model for NbV. *J. Chem. Soc., Dalton Trans., 20*, 3055–3059. http://dx.doi.org/10.1039/DT9940003055 (Etxebarria 1994).

[7] Fisher, A. & Szabó, Z. (2004). Sodium penta-fluoro-uranate(IV) monohydrate, Na[UF_5]·H_2O. *Acta Crystallographica, Section E., 30*, i45-i46. https://doi.org/10.1107/S1600536804003241 Online 20 February 2004. (Fisher 2004).

[8] Gorbaty, Y. E. & Kalinichev, A. G. (1995). Hydrogen bonding in supercritical water. 1. Experimental results. *J. Phys. Chem., 99*, 5336–5340. DOI: 10.1021/j100015a016 Accessed April 1995. (Gorbaty 1995).

[9] Grenthe, Ingmar., Drożdżyński, Janusz., Fujino, Takeo., Buck, Edgar C., Albrecht-Schmitt, Thomas E. & Wolf, Stephen F. (2010). Uranium. 1 (5): 253-698. In: *The Chemistry of the Actinide and Transactinide Elements*. 4th edition. (Morss, Lester R., Edelsteine Norman M., and Fuger Jean eds.). Springer, Netherlands., 4521 p. (Grenthe 2010).

[10] Helgeson, Harold C. & Kirkham, David H. (1974). Theoretical prediction of the thermodynamic behavior of aqueous electrolytes at high pressures and temperatures. I. Summary of the thermodynamic/electrostatic properties of the solvent. *Am. J. Sci., 274*, 1089–1098. DOI: 10.2475/ajs.274.10.1089. Online December 1974. (Helgeson 1974_1).

[11] Helgeson, Harold C. & Kirkham, David H. (1974). Theoretical prediction of the thermodynamic behavior of aqueous electrolytes at high pressures and temperatures. II. Debye-Huckel parameters for activity coefficients and relative partial molal properties. *Am. J. Sci., 274*, 1199-1261. DOI: 10.2475/ajs.274.10.1199. Online December 1, 1974. (Helgeson 1974_2).

[12] IAEA-TECDOC-1629. (2009). *World Distribution of Uranium Deposits (UDEPO) with Uranium Deposit Classification*. IAEA-TECDOC-1629, International Atomic Energy Agency, October 2009. Vienna, Austria., 126 p. http://www-pub.iaea.org/MTCD/publications/PDF/TE_1629_web.pdf. (IAEA 2009).

[13] Ilton, Eugene S. & Bagus, Paul S. (2011). XPS determination of uranium oxidation states. *Surf. Interface Anal.*, *43*, 1549–1560. DOI 10.1002/sia.3836 Accessed on line 5 Oct 2011 (Ilton 2011).

[14] Johnson, J. W., Oelkers, E. H. & Helgeson, H. C. (1992). SUPCRT92: A software package for calculating the standard molal thermodynamic properties of minerals, gases, aqueous species, and reactions from 1 to 5000 bar and 0 to 1000°C. *Computers Geosci.*, *18*, 899–947. http://dx.doi.org/10.1016/0098-3004(92)90029-Q. Online January, 1992. (Johnson 1992).

[15] Kalinichev, Andrey G. (2017). Universality of hydrogen bond distributions in liquid and supercritical water. *J. Mol. Liq.*, *241*, 1038–1043. http:// dx.doi.org/ 10.1016/ j.molliq.2017.06.114. Available online 27 June 2017. (Kalinichev 2017).

[16] Kalinichev, A. G. & Bass, J. D. (1994). Hydrogen bonding in supercritical water: a Monte Carlo simulation. *Chem. Phys. Lett.*, *231*, 301–307. http://dx.doi.org/10.1016/0009-2614(94)01245-8. Published 23 Dec. 1994. (Kalinichev 1994).

[17] Kalinichev, A. G. & Churakov, S. V. (1999). Size and topology of molecular clusters in supercritical water: a molecular dynamics simulation. *Chem. Phys. Lett.*, *302*, 411–417. https://doi.org/10.1016/S0009-2614(99)00174-8. Available online 31 March 1999. (Kalinichev 1999).

[18] Keppler, Hans. & Wyllie, Petter J. (1990). Role of fluids in transport and fractionation of uranium and thorium in magmatic processes. *Nature*, *348*, 531–533. doi:10.1038/348531a0. 06 Accessed December 1990. (Keppler 1990).

[19] Keppler, Hans. & Wyllie, Petter J. (1991). Partitioning of Cu, Sn, Mo, W, U, and Th between melt and aqueous fluid in the systems haplogranite-H_2O−HCl and haplogranite-H_2O−HF. *Contributions to Mineralogy and Petrology.*, *109*, 139–150. https://doi.org/10.1007/BF00306474. Accessed December 1991. (Keppler 1991).

[20] Kotelnikova, Z. A. & Kotelnikov, A. R. (2008). NaF-bearing fluids: experimental investigation at 500-800°C and P = 2000 bar using synthetic fluid inclusions in quartz. *Geochemistry International*, *46*,

48-61. DOI: 10.1134/S0016702908010047. Accessed January 2008. (Kotelnikova 2008).

[21] Kovalenko, V. I. (1977). *Petrology and Geochemistry of Rare-Metal Granitoids*. Novosibirsk: Nauka SO. 207 p. (in Russian). (Kovalenko 1977).

[22] Kovalenko, N. I., Ryzhenko, B. N., Prisyagina, N. I. & Bychkova, Ya. V. (2011). Experimental study of uraninite solubility in aqueous HCl solutions at 500°C and 1 kbar. *Geochemistry International*, *49*, 262–273. DOI: 10.1134/S0016702911010058. Available online 23 March 2011. (Kovalenko 2011).

[23] Kovalenko, N. I., Ryzhenko, B. N., Prisyagina, N. I. & Bychkova, Ya. V. (2012). Experimental determination of uranium (IV) speciation in HF solutions at 500°C and 1000 bar. *Geochemistry International*, *50*, 18–25. DOI: 10.1134/S0016702912010053. Available online 30 December 2011. (Kovalenko 1012).

[24] Lukyanova, E. V., Akinfiev, N. N., Zotov, A. V., Rass, I. T., Kotova, N. P. & Korzhinskaya, V. S. (2017). Niobium in hydrothermal systems related to alkali granites: Thermodynamic description of hydroxo and hydroxofluoride complexes. *Geology of Ore Deposits*, *59*, 305–314. DOI: 10.1134/S1075701517040031. Available online 03 September 2017. (Lukyanova 2017).

[25] Martin, H., Albers, A. & Dust, H. P. (1951). Doppelfluoride des Uranhexafluorids. *Z. anorg. u. allgem. Chem.*, *265*, 128-138. DOI: 10.1002/zaac.19512650114. Published June 1951. (Martin 1951).

[26] Naumov, G. B. (1959). To the question of carbonate form of uranium transportation in hydrothermal solution. *Geochemistry.*, *1*, 6-19. (in Russian). (Naumov 1959).

[27] Naumov, G. B. (1978). *Principles of physicochemical modelling of uranium ore formation*. Moscow: Atomizdat., 213 p. (in Russian). (Naumov 1978).

[28] Parks, George A. & Pohl, Demetrius S. (1988). Hydrothermal solubility of uraninite. *Geochim. Cosmochim. Acta*, *52*, 863-875. https://doi.org/10.1016/0016-7037(88)90357-2. Accepted 19 January 1988. (Parks 1988).

[29] Peiffert, Chantal., Nguyen-Ttung, Chinh. & Cuney, Michel. (1996). Uranium in granitic magmas: Part 2. Experimental determination of uranium solubility and fluid melt partition coefficients in the uranium oxide–haplogranite–H_2O–NaX (X = Cl, F) system at 770°C, 2 kb. *Geochim. Cosmochim. Acta.*, *60*, 1515–1529. https://doi.org/10.1016/0016-7037(96)00039-7. Accepted 24 January 1996. (Peiffert 1996).

[30] Peiffert, C., Nguyen-Trung, Ch., Palmer, D. A., Laval, J. P. & Giffaut, E. (2010). Solubility of B–Nb_2O_5 and the hydrolysis of niobium(V) in aqueous solution as function of temperature and ionic strength. *J. Solution Chem*, *39*, 197–218. DOI 10.1007/s10953-010-9495-z. Available online 2 March 2010. (Peiffert 2010).

[31] Rafalsky, R. P. (1973). *Hydrothermal equilibria and processes of minerals formation.* Moscow: Nauka., 288 p. Free book site: http://www.geokniga.org/books/7820. (Rafalsky 1973).

[32] Redkin, A. F. (1996). "Influence of oxygen fugacity and solution composition on uranium oxides solubility at temperatures to 600°C and P=1 kbar (by experimental and thermodynamic data)". *Extended Abstracts of 4th International Conference on Nuclear and Radiochemistry NRC4*, St-Malo, France, September 8-13, *2*, E-P61. (Redkin 1996).

[33] Redkin, A. F. (2013). Study of solubility of UO_2 and U(VI) hydroxides in water at high temperature and pressure. In: *Materials of the scientific meeting dedicated scientific activities of Prof. I.L. Khodakovsky.* Dubna., P. 73-79 (in Russian) (Redkin 2013_2).

[34] Redkin, A. F. & Borodulin, G. P. (2010). Pyrochlores as indicators of the uranium potential of magmatic melts. *Doklady Earth Sciences*, *432* (2), 787-790. DOI: 10.1134/S1028334X10060176. June 2010. (Redkin 2010).

[35] Redkin, Alexander F., Ionov, Andrey M. & Kotova, Nataliya P. (2013). Hydrothermal synthesis of pyrochlores and their characterization. *Phys. Chem. Minerals*, *40*, 733–745. DOI: 10.1007/s00269-013-0608-2. Available online October 2013. (Redkin 2013_1).

[36] Redkin, Alexander F., Kotova, Nataliya P. & Shapovalov, Yury B. (2015). Liquid immiscibility in the system NaF–H_2O at 800°C and 200–230 MPa and its effect on the microlite solubility. *J. Solution Chem.*, *44*, 2008-2026. DOI: 10.1007/s10953-015-0394-1. Available online 15 October 2015 (Redkin 2015_1).

[37] Redkin, A. F., Kotova, N. P. & Shapovalov, Yu. B. (2015). "Effect of liquid immiscibility on microlite solubility in the system NaF-H_2O at 800°C". *Materials of a conference: Deposits of strategic metals: patterns of location, sources of matter, conditions and mechanisms of formation. All-Russian conference dedicated 85th Anniversary of IGEM RAS foundation.* Moscow: IGEM RAS., P. 144-145 (in Russian). (Redkin 2015_2).

[38] Redkin, A. F., Kotova, N. P. & Shapovalov, Yu. B. (2016). Liquid immiscibility in the system NaF–H_2O and microlite solubility at 800°C. *Doklady Earth Sciences*, *469*(1), 722-727. DOI: 10.1134/S1028334X16070151. Received March 29, 2016 (Redkin 2016_1).

[39] Redkin, A. F. & Velichkin, V. I. (2013). Uranium fluorides in hydrothermal-magmatic systems. *Doklady Earth Sciences*, *450* (1), 544–547. DOI: 10.1134/S1028334X13050218 (Redkin 2013_2).

[40] Redkin, A. F. & Kotova, N. P. (2016). "Effect of NaF concentration on pyrochlore solubility at 800°C and 200 MPa." *In proceedings of XVII international conference "Physical-chemical and petrophysical researches in the Earth's sciences*. Moscow., Sept. 26-28, 2016, P. 290-292. (Redkin 2016_2).

[41] Redkin, A. F., Savelyeva, N. I., Sergeyeva, E. I., Omelyanenko, B. I., Ivanov, I. P. & Khodakovsky, I. L. (1989). Investigation of uraninite UO_2 solubility under hydrothermal conditions. *Sci. Geol. Bull. Strasbourg*, *42*(4), 329-334. (Redkin 1989).

[42] Redkin, A. F., Velichkin, V. I., Aleshin, A. P. & Borodulin, G. P. (2009). Interaction of model F-bearing silicic melt with chloride fluid, uraninite, and columbite at 750°C and 1000–2000 bar and its implications for estimation of the ore-forming capability of the upper crustal magma chamber beneath the Strel'tsovka caldera, Eastern Transbaikalia. *Geology of Ore Deposits*, *51*, 290–304. Received

February 20, 2008. DOI: 10.1134/S1075701509040035. (Redkin 2009).

[43] Schindler, M., Hawthorne, F. C., Freund, M. S. & Burns, P. C. (2009). XPS spectra of uranyl minerals and synthetic uranyl compounds. I: The U 4f spectrum. *Geochim. Cosmochim. Acta*, *73*, 2471−2487. https://doi.org/10.1016/j.gca.2008.10.042. Accepted 30 October 2008. (Schindler 2009).

[44] Shock, Everett L., Oelkers, Eric H., Johnson, James W., Sverjensky, Dimitri A. & Helgeson, Harold C. (1992). Calculation of the thermodynamic properties of aqueous species at high pressures and temperatures: Effective electrostatic radii, dissociation constants and standard partial molal properties to 1000°C and 5 kbar. *J. Chem. Soc., Faraday Trans.*, *88*, 803-826. http://dx.doi.org/10.1039/FT9928800803. (Shock 1992).

[45] Shock Everett, L., Sassani, David C., Willis, Marc. & Sverjensky, Dimitri A. (1997). Inorganic species in geologic fluids: Correlations among standard molal thermodynamic properties of aqueous ions and hydroxide complexes. *Geochim. Cosmochim. Acta*, *61*, 907-950. Available online 4 June 1998. https://doi.org/10.1016/S0016-7037(96)00339-0 (Shock 1997_1).

[46] Shock, Everett L., Sassani, David C. & Betz, Heidi. (1997). Uranium in geological fluids-estimates of standard partial molal properties, oxidation potentials, and hydrolysis constants at high temperatures and pressures. – Review. *Geochim. Cosmochim. Acta*, *61*, 4245-4266. Accepted 15 June 1997. https://doi.org/10.1016/S0016-7037(97)00240-8. (Shock 1997_2).

[47] Shvarov, Yuri. (2008). HCh: new potentialities for the thermodynamic simulation of geochemical systems offered by Windows. *Geochemistry International*, *46*, 834–839. Received April 22, 2007. http://dx.doi.org/10.1134/s0016702908080089. (Shvarov 2008).

[48] Shvarov, Yuri. (2015). A suite of programs, OptimA, OptimB, OptimC, and OptimS compatible with the Unitherm database, for deriving the thermodynamic properties of aqueous species from

solubility, potentiometry and spectroscopy measurements. *Applied Geochemistry*, *55*, 17-27. http:// dx.doi.org/ 10.1016/ j.apgeochem. 2014.11.021. (Shvarov 2015).

[49] Shvarov, Yuri. & Bastrakov, Eugeniy. (1999). HCh: a software package for geochemical equilibrium modelling. *User's Guide. Australian Geological Survey Organization. Record 1999/25.* https://d28rz98at9flks.cloudfront.net/25473/Rec1999_025.pdf. (Shvarov 1999).

[50] Timofeev, A., Migdisov, Art A. & Williams-Jones, A. E. (2015). An experimental study of the solubility and speciation of niobium in fluoride-bearing aqueous solutions at elevated temperature. *Geochim. Cosmochim. Acta*, *158*, 103–111. Accepted 15 February 2015. http://dx.doi.org/10.1016/j.gca.2015.02.015. (Timofeev 2015).

[51] Timofeev, A., Migdisov, Art A., Williams-Jones, A. E. (2017). An experimental study of the solubility and speciation of tantalum in fluoride-bearing aqueous solutions at elevated temperature. *Geochim. Cosmochim. Acta*, *197*, 294–304. Accepted 19 October 2016. http://dx.doi.org/10.1016/j.gca.2016.10.027. (Timofeev 2017).

[52] Thoma, R. E. & Grimes, W. R. (1957). *Phase equilibrium diagrams for fused salt systems*. AEC research and development report. ORNL-2295. Reactors-Special Feature of Aircraft Reactors. Oak Ridge Tennessee., Issued June 24, 1957, 98 p. (p.23 system NaF-UF_4). (Thoma 1957).

[53] Tropper, P. & Manning, C. E. (2007). The solubility of fluorite in H_2O and H_2O–NaCl at high pressure and temperature. *Chemical Geology*, *242*, 299–306. Accepted 27 January 2007. doi: 10.1016/ j.chemgeo.2007.03.017. (Tropper 2007).

[54] Yeon, Jeongho., Smith, Mark D., Morrison, Gregory. & zur, Loye Hans-Conrad. (2015). Trivalent Cation-Controlled Phase Space of New U(IV) Fluorides, $Na_3MU_6F_{30}$ (M = Al^{3+}, Ga^{3+}, Ti^{3+}, V^{3+}, Cr^{3+}, Fe^{3+}): Mild Hydrothermal Synthesis Including an in Situ Reduction Step, Structures, Optical, and Magnetic Properties. *Inorg. Chem.*, *54*, 2058–2066. Publication Date February 5, 2015 DOI: 10.1021/ ic503034t. (Yeon 2015).

[55] Zaraisky, G. P., Aksyuk, A. M., Devyatova, V. N., Udoratina, O. V. & Chevychelov, V. Yu. (2009). The Zr/Hf ratio as a fractionation indicator of rare-metal granites. *Petrology*, *17*, 25–45. Received June 11, 2008. doi:10.1134/S0869591109010020. (Zaraisky 2009).

[56] Zaraisky, George P., Korzhinskaya, Valentina. & Kotova, Nataliya. (2010). Experimental studies of Ta_2O_5 and columbite–tantalite solubility in fluoride solutions from 300 to 550°C and 50 to 100 MPa. *Miner. Petrol.*, *99*, 287–300. Accepted: 23 February 2010. DOI 10.1007/s00710-010-0112-z. (Zaraisky 2010).

[57] Zaraisky, G. P., Chevychelov, V. Yu., Aksyuk, A. M., Korzhinskaya, V. S., Kotova, N. P., Redkin, A. F. & Borodylin, G. P. (2008). Collected papers "*Experimental studies of endogenic processes*". Chernogolovka., P. 86-109 [in Russian]. (Zaraisky 2008).

[58] Zharikov, V. A., Ivanov, I. P., Omel'yanenko, B. I., Red'kin, A. F. & Yudintsev, S. V. (1987). Experimental Study of the Solubility of Uraninite Granitic Melts and fluid Solutions at High pressure and Temperatures. *International Geology Review.*, *29*, 997-1004. Published online 29 Jun 2010. http://dx.doi.org/ 10.1080/ 0020 6818709466194. (Zharikov 1987).

[59] Veksler, Ilia V. (2004). Liquid immiscibility and its role at the magmatic-hydrothermal transition: a summary of experimental studies. *Chemical Geology*, *210*, 7—31. Available online 28 July 2004. https://doi.org/10.1016/j.chemgeo.2004.06.002. (Veksler 2004).

[60] Wood, Scott A. (2005). *The aqueous geochemistry of zirconium, hafnium, niobium and tantalum.* In: Linnen R.L., and Samson I.M., eds. *Rare-Element Geochemistry and Mineral Deposits: Geological Association of Canada, GAC Short Course Notes*, 17, p. 217–268. (Wood 2005).

In: Solution Chemistry
Editor: Yongliang Xiong

ISBN: 978-1-53613-101-7

Chapter 2

EXPERIMENTAL DETERMINATION OF BRUCITE SOLUBILITY IN NACL SOLUTIONS AT ELEVATED TEMPERATURES

Leslie Kirkes and Yongliang Xiong*
Sandia National Laboratories (SNL), Carlsbad, NM, US

ABSTRACT

Salt formations have been recommended for nuclear waste isolation since the 1950's by the U.S. National Academy of Science. This recommendation has been implemented in southeast New Mexico where the Waste Isolation Pilot Plant (WIPP) has been built to isolate defense-related transuranic waste. The WIPP is located in a bedded salt formation, the Salado Formation. Placement of crystalline MgO, which hydrates rapidly to form brucite, is the only engineered barrier employed in the WIPP design. The MgO acts as a chemical conditioner in the WIPP repository in controlling the fugacity of carbon dioxide. Similarly, an $Mg(OH)_2$-based engineered barrier is proposed for the German Asse salt mine repository. Thus, the solubility of brucite is of interest to salt repository programs which can expect a variety of temperatures within

* Corresponding Author Email: yxiong@sandia.gov.

the repository and a variety of fluids (brines) coming in contact with the waste.

Salt repository programs are not the only programs that stand to benefit from the information presented in this book chapter. There are other applications where this information is of interest. In natural environments brucite frequently precipitates from hyperalkaline hydrothermal fluids with high ionic strengths. For instance, brucite chimneys have been observed to form at elevated temperatures in ocean floors. The information presented in this work can be used to accurately model the formation of such brucite chimneys.

In this study, we have determined solubilities of brucite as a function of ionic strength in NaCl solutions to I = 5.6 mol•kg^{-1} at elevated temperatures to 353.15 K. In our solubility measurements, we first independently determined the correction factors for converting pH readings to pH_m (negative logarithm of hydrogen ion concentration on a molal scale, mol•kg^{-1}) in NaCl solutions from 0.01 to 5.6 mol•kg^{-1} at elevated temperatures. Using the SIT model, we obtain the solubility constants for brucite at infinite dilution as a function of temperature,

$$Mg(OH)_2(cr) + 2H^+ \rightleftharpoons Mg^{2+} + 2H_2O(l)$$

which can be described by the following expression,

$$\log_{10} K^0_{s2} = \frac{4306.8369}{T} + 2.6383$$

where T is temperature in K. This expression can be used from 273.15 K to 373.15 K.

Keywords: nuclear waste management, hydrolysis, Specific Ion Interaction Theory (SIT), Salt Formations, High Level Nuclear Waste, Waste Isolation Pilot Plant (WIPP), Brucite chimneys

INTRODUCTION

Salt formations are ideal for nuclear waste isolation, as noted by the National Academy of Science as early as the 1950's (U.S. National Academy of Science, 1957). There are already two repositories in salt formations in the world. The Waste Isolation Pilot Plant (WIPP) is a U.S.

Department of Energy (DOE) repository for disposal of defense transuranic waste. The WIPP is located in a bedded salt formation and is located in southeast New Mexico, USA. The German Asse repository is located in a domal salt formation (Schűessler et al., 2001).

In WIPP and Asse, brucite, a hydrated magnesium oxide mineral, influences the near-field geochemistry. This is because the WIPP will use crystalline MgO and Asse will use hydrated MgO emplaced alongside the waste as an engineered barrier (e.g., Krumhansl et al., 2000; Xiong and Snider, 2008). MgO has been placed in the WIPP since operations began. An $Mg(OH)_2$-based engineered barrier is also proposed for the German Asse salt mine repository (Schűessler et al., 2001). Crystalline MgO hydrates rapidly forming brucite, which based on its solubility, conditions the near-field environment and controls the fugacity of carbon dioxide. Therefore, brucite solubilities in high ionic strength solutions at various temperatures are required to predict the near-field geochemical environment both for WIPP and Asse. Of note is that the German Asse repository is expected to witness elevated temperatures and both repositories have high ionic strength brines present.

In addition, there are several candidate repositories in salt formations, including Gorleben (Schwartz, 2012), Morsleben (Behlau and Mingerzahn, 2001; Brewitz et al., 2008) in Germany, and the sites in the Mid North Sea High and the East Midlands Shelf in UK (Stewart, 2002). Although the engineered barrier for those candidate repositories has not been decided, MgO- or $Mg(OH)_2$-based engineered barrier would be a strong candidate.

The solubility of brucite as a function of ionic strength at room temperature has been well characterized (e.g., Altmaier et al., 2003; Xiong, 2003, 2008). However, the solubilities of brucite at elevated temperatures in high ionic strength solutions, which may be relevant to disposal of heat-generating waste in salt formations is not well known.

In addition, brucite has been observed to form in saline, hyperalkaline, hydrothermal fluids (Monnin et al., 2014; Okumura et al., 2016). In these environments, the temperatures are elevated up to 90°C, and the fluids are of high ionic strength, and brucite supports unique ecosystems with dense *Phyllochaetopterus* (e.g., Okumura et al., 2016). Therefore, brucite solubilities at elevated temperatures in high ionic strength solutions contribute to modeling the formation of brucite in those environments.

The objective of this work is to experimentally determine solubilities of brucite as a function of ionic strength in NaCl solutions up to 5.6 mol•kg^{-1} at elevated temperatures. In our work, the correction factors that are used to convert pH readings to pcH (negative logarithm of hydrogen ion concentration on a molar scale) are separately determined.

METHODS

All materials (e.g., NaCl, $Mg(OH)_2$) used in this study are reagent grade from Fisher Scientific. Deionized (DI) water with 18.3 MΩ was produced by a *PURELAB Classic Water System* from ELGA. Degassed DI water was used for preparation of all starting solutions. The degassed DI water was obtained by bubbling high purity argon gas (purity 99.996%) from Matheson Tri-Gas, Inc., through DI water for at least one hour, following a procedure similar to that described by Wood et al., (2002). Starting solutions were prepared such that the equilibrium solubility was approached from undersaturation with respect to brucite. The starting solutions included NaCl solutions ranging from 0.010 to 5.6 mol•kg^{-1} NaCl.

The experiments were conducted at five temperatures from 40 ± 0.05°C through 80 ± 0.05°C at an increment of 10°C. For each of the experiments undersaturated with respect to brucite, about 1 gram of $Mg(OH)_2$ (cr) was placed into a 150 mL high density polyethylene plastic bottle containing 100 mL of starting solution.

At specific intervals, the pH reading of each experimental solution was measured with an Orion-Ross combination pH glass electrode. Before each measurement, the pH meter was calibrated with three pH buffers (pH 4, pH 7 and pH 10) at the experimental temperature. For solutions with ionic strengths higher than 0.10 mol•kg^{-1}, the observed solution pH readings were converted to negative logarithm of hydrogen-ion concentrations (pcH) on a molar scale using a correction factor, *A* (see below).

The relation between the pH electrode reading (pH_{ob}) and pcH can be expressed as (Xiong et al., 2010):

$$pcH = pH_{ob} + A_M \tag{1}$$

where A_M is a correction factor on a molar scale. The correction factors were determined at the experimental temperatures as a function of ionic strength by following a procedure similar to that of Xiong et al., (2010). Then, pcH's were converted to negative logarithms of hydrogen-ion concentrations (pH_m) on a molal scale according to the following equation (Xiong et al., 2010),

$$pH_m = pH_{ob} + A_m = pH_{ob} + A_M - \log \Theta \tag{2}$$

where A_m is a correction factor on a molal scale, and Θ is a conversion factor from molality to molarity, which is computed from the density of a NaCl solution at the concentration and temperature of interest from Sőhnel and Novotný (1985).

After a pH measurement was taken, certain amounts of solution were withdrawn from an experimental run. The solution was filtered using a 0.2 μm syringe filter. The filtered solution was then weighed and acidified with 0.5 mL of concentrated HNO_3 (TraceMetal grade from Fisher Scientific). The extracted solution was then diluted to 10 mL with DI water so that dissolved ΣMg^{2+} and Na^+ concentrations could be determined.

Elemental concentrations of magnesium were determined by using the Perkin Elmer Optima 8300 inductively coupled plasma atomic emission spectrometer (ICP-AES). Calibration blanks and standards were precisely matched with experimental matrices. The correlation coefficients of calibration curves in all measurements were better than 0.9995. The analytical precision is better than 1.00% in terms of the relative standard deviation (RSD) based on replicate analyses.

Solid phase identification was performed by using a Bruker AXS, Inc., D8 Advance X-ray diffractometer (XRD) with a Sol-X detector. XRD patterns were collected using CuKα radiation at a scanning rate of 1.33°/min for a 2θ range of 10–90°. There is no new phase observed.

Scanning electron microscope (SEM) analyses of the starting material, $Mg(OH)_2(cr)$, indicate that the majority of particles are larger than 1 μm (Figure 1). The particle size of the solubility-controlling phase, brucite, after the experiments, is similar to that of the starting material (Figures 2 and 3). The particle size of brucite in the experiments performed in this work is similar to that observed in Xiong (2008). Accordingly, based on the assessment in Xiong (2008), the surface energy contribution to the solubility of brucite is insignificant in this study.

Figure 1. Scanning electron microscope (SEM) image of the starting material, $Mg(OH)_2(cr)$.

Figure 2. Scanning electron microscope (SEM) image of brucite, $Mg(OH)_2(cr)$, in 1.0 mol•kg^{-1} NaCl solution at 80°C, after the experiment.

Figure 3. Scanning electron microscope (SEM) image of brucite, $Mg(OH)_2(cr)$, in 5.6 mol•kg^{-1} NaCl solution at 80°C, after the experiment.

EXPERIMENTAL RESULTS AND THERMODYNAMIC CALCULATIONS

Correction Factors at Elevated Temperatures as a Function of Ionic Strength

In Table 1, the experimentally determined correction factors in NaCl solutions at various temperatures as a function of ionic strength are listed. In addition, the correlations between correction factors and molalities of NaCl are presented as linear equations at each temperature in Table 2, based on the experimental data from Table 1.

Solubility of Brucite in NaCl Solutions at Elevated Temperatures

In Table 3, the raw experimental data from this study are tabulated. The total magnesium concentrations are the result of the following reactions:

$$Mg(OH)_2(cr) + 2H^+ \rightleftharpoons Mg^{2+} + 2H_2O(l) \quad (3)$$

$$Mg^{2+} + H_2O(l) \rightleftharpoons MgOH^+ + H^+ \quad (4)$$

Table 1. Experimentally determined correction factors at different temperatures as a function of ionic strength

Temperature, °C	NaCl, mol•kg^{-1}	A_M	A_m
40	1.0	0.007	–0.00469
	2.1	0.191	0.170
	3.2	0.363	0.332
	4.4	0.564	0.522
	5.6	0.762	0.708
50	1.0	–0.01	–0.0237
	2.1	0.237	0.214
	3.2	0.372	0.338
	4.4	0.579	0.535
	5.6	0.750	0.695
60	1.0	0.08	0.0641
	2.1	0.218	0.192
	3.2	0.433	0.398
	4.4	0.637	0.590
	5.6	0.758	0.700
70	1.0	0.121	0.105
	2.1	0.238	0.212
	3.2	0.429	0.394
	4.4	0.597	0.550
	5.6	0.770	0.712
80	1.0	0.124	0.103
	2.1	0.251	0.221
	3.2	0.457	0.417
	4.4	0.615	0.564
	5.6	0.713	0.651

Table 2. Concentration dependence of correction factors at constant temperatures

T, °C	A_M as a function of molality	A_m as a function of molality
40	$A_M = -0.1565 + 0.1638\ m_{NaCl}$	$A_m = -0.1587 + 0.1547\ m_{NaCl}$
50	$A_M = -0.1417 + 0.1618\ m_{NaCl}$	$A_m = -0.1462 + 0.1527\ m_{NaCl}$
60	$A_M = -0.07746 + 0.1542\ m_{NaCl}$	$A_m = -0.08406 + 0.1451\ m_{NaCl}$
70	$A_M = -0.03886 + 0.1441\ m_{NaCl}$	$A_m = -0.04546 + 0.1351\ m_{NaCl}$
80	$A_M = -0.004252 + 0.1338\ m_{NaCl}$	$A_m = -0.01614 + 0.1249\ m_{NaCl}$

In this work, we compute the equilibrium quotients for Reaction (3), Q_{s2}, based on our solubility data. The free magnesium concentrations, $m_{Mg^{2+}}$, are calculated from total magnesium concentrations after correcting contributions from $MgOH^+$, m_{MgOH^+}, according to the equilibrium quotients for Reaction (4), $Q_{1,1}$, from Palmer and Wesolowski (1997) in NaCl solutions at the respective temperatures and ionic strengths:

$$m_{\Sigma Mg} = m_{Mg^{2+}} + m_{MgOH^+} \tag{5}$$

$$Q_{1,1} = \frac{m_{MgOH^+} \times m_{H^+}}{m_{Mg^{2+}}} = \frac{(m_{\Sigma Mg} - m_{Mg^{2+}}) \times m_{H^+}}{m_{Mg^{2+}}} \tag{6}$$

Therefore, using the experimentally determined total magnesium and hydrogen ion concentrations, free magnesium concentrations can be determined by solving Eq. (6) for $m_{Mg^{2+}}$. After obtaining $m_{Mg^{2+}}$, equilibrium quotients for Reaction (3) can be computed according to the following equation:

$$Q_{s2} = \frac{m_{Mg^{2+}}}{(m_{H^+})^2} \tag{7}$$

Table 3. Solubilities of brucite in NaCl solutions at elevated temperatures determined in this work

Exp. #	T, C	NaCl, mol•kg^{-1}	Exp. time, d	pH	pH_m	$m_{\Sigma Mg}$, mol•kg^{-1}
Brucite-0.01-1 40C	40	0.01	112	9.71	9.71	1.48E-03
			209	9.67	9.67	1.54E-03
			344	9.66	9.66	1.72E-03
Brucite-0.01-2 40C	40	0.01	112	9.75	9.75	1.48E-03
			209	9.69	9.69	1.59E-03
			344	9.65	9.65	1.88E-03
Brucite-0.1-1 40C	40	0.1	112	9.81	9.81	1.69E-03
			211	9.81	9.81	1.67E-03
			344	9.76	9.76	1.90E-03
Brucite-0.1-2 40C	40	0.1	112	9.82	9.82	1.76E-03
			211	9.85	9.85	1.70E-03
			344	9.76	9.79	1.89E-03
Brucite-1.0-1 40C	40	1.0	112	9.82	9.82	2.47E-03
			211	9.70	9.70	2.99E-03
			344	9.74	9.74	3.37E-03
Brucite-1.0-2 40C	40	1.0	112	9.78	9.78	2.35E-03
			211	9.73	9.73	2.39E-03
			344	9.77	9.77	2.65E-03
Brucite-2.0-1 40C	40	2.1	112	9.68	9.85	2.57E-03
			211	9.63	9.80	2.54E-03
			344	9.69	9.86	2.75E-03
Brucite-2.0-2 40C	40	2.1	112	9.67	9.84	2.58E-03
			211	9.64	9.81	2.44E-03
			344	9.68	9.85	2.68E-03
Brucite-3.0-1 40C	40	3.2	112	9.44	9.77	2.61E-03
			211	9.46	9.79	2.61E-03
			344	9.51	9.84	3.22E-03
Brucite-3.0-2 40C	40	3.2	112	9.41	9.74	2.58E-03
			211	9.39	9.72	3.44E-03
			344	9.45	9.78	4.31E-03

Exp. #	**T, C**	**NaCl, mol•kg^{-1}**	**Exp. time, d**	**pH**	**pH$_m$**	**$m_{\Sigma Mg}$, mol•kg^{-1}**
Brucite-4.0-1 40C	40	4.4	112	9.27	9.79	2.45E-03
			211	9.32	9.84	2.91E-03
			344	9.33	9.85	4.12E-03
Brucite-4.0-2 40C	40	4.4	112	9.30	9.82	2.44E-03
			211	9.36	9.88	2.47E-03
			344	9.41	9.93	2.68E-03
Brucite-5.0-1 40C	40	5.6	112	9.18	9.89	2.30E-03
			211	9.19	9.90	2.26E-03
			344	9.25	9.96	2.95E-03
Brucite-5.0-2 40C	40	5.6	112	9.19	9.90	2.30E-03
			211	9.19	9.90	2.51E-03
			344	9.24	9.95	2.79E-03
Brucite-0.01-1 50C	50	0.01	111	9.43	9.43	1.63E-03
			205	9.50	9.50	1.94E-03
			343	9.45	9.45	2.07E-03
Brucite-0.01-2 50C	50	0.01	111	9.51	9.51	1.64E-03
			205	9.45	9.45	1.66E-03
			343	9.49	9.49	1.73E-03
Brucite-0.1-1 50C	50	0.1	111	9.70	9.70	1.84E-03
			205	9.52	9.52	1.95E-03
			343	9.54	9.54	2.17E-03
Brucite-0.1-2 50C	50	0.1	111	9.72	9.72	1.85E-03
			205	9.53	9.53	1.80E-03
			343	9.57	9.57	1.92E-03
Brucite-1.0-1 50C	50	1.0	111	9.50	9.48	2.57E-03
			205	9.50	9.48	2.51E-03
			343	9.56	9.54	2.66E-03
Brucite-1.0-2 50C	50	1.0	111	9.55	9.53	2.62E-03
			205	9.48	9.46	2.61E-03
			343	9.57	9.55	2.78E-03
Brucite-2.0-1 50C	50	2.1	111	9.26	9.47	2.84E-03
			205	9.37	9.59	2.81E-03
			343	9.42	9.64	3.03E-03

Table 3. (Continued)

Exp. #	T, C	NaCl, mol•kg^{-1}	Exp. time, d	pH	pH$_m$	$m_{\Sigma Mg}$, mol•kg^{-1}
Brucite-2.0-2 50C	50	2.1	111	9.39	9.60	2.66E-03
			205	9.38	9.59	2.74E-03
			343	9.49	9.70	2.79E-03
Brucite-3.0-1 50C	50	3.2	112	9.07	9.41	2.80E-03
			205	9.26	9.60	2.58E-03
			343	9.35	9.69	2.79E-03
Brucite-3.0-2 50C	50	3.2	112	8.98	9.32	2.67E-03
			205	9.23	9.57	2.59E-03
			343	9.36	9.70	2.69E-03
Brucite-4.0-1 50C	50	4.4	112	9.01	9.54	2.58E-03
			205	9.10	9.63	2.44E-03
			343	9.11	9.64	2.59E-03
Brucite-4.0-2 50C	50	4.4	112	8.98	9.51	2.72E-03
			205	9.08	9.61	2.49E-03
			343	9.18	9.71	3.04E-03
Brucite-5.0-1 50C	50	5.6	112	8.89	9.58	2.40E-03
			205	8.97	9.66	2.27E-03
			343	9.09	9.78	2.49E-03
Brucite-5.0-2 50C	50	5.6	112	8.89	9.58	2.45E-03
			205	8.97	9.66	2.35E-03
			343	9.09	9.78	2.51E-03
Brucite-0.01-1 60C	60	0.01	105	9.22	9.22	1.60E-03
			127	9.30	9.30	1.61E-03
			342	9.22	9.22	2.09E-03
Brucite-0.01-2 60C	60	0.01	105	9.23	9.23	1.65E-03
			127	9.31	9.31	1.61E-03
			342	9.32	9.32	1.80E-03
Brucite-0.1-1 60C	60	0.1	105	9.34	9.34	2.00E-03
			127	9.25	9.25	1.97E-03
			342	9.33	9.54	2.21E-03

Exp. #	T, C	NaCl, mol•kg^{-1}	Exp. time, d	pH	pH_m	$m_{\Sigma Mg}$, mol•kg^{-1}
Brucite-0.1-2 60C	60	0.1	105	9.30	9.30	1.87E-03
			127	9.27	9.27	1.85E-03
			342	9.33	9.33	1.97E-03
Brucite-1.0-1 60C	60	1.0	105	9.26	9.32	2.48E-03
			127	9.27	9.33	2.47E-03
			342	9.37	9.43	2.63E-03
Brucite-1.0-2 60C	60	1.0	105	9.23	9.29	2.56E-03
			127	9.28	9.34	2.47E-03
			342	9.37	9.43	2.65E-03
Brucite-2.0-1 60C	60	2.1	105	9.13	9.32	2.72E-03
			127	9.18	9.37	2.64E-03
			342	9.27	9.46	2.75E-03
Brucite-2.0-2 60C	60	2.1	105	9.12	9.31	2.67E-03
			127	9.17	9.36	2.57E-03
			342	9.27	9.46	2.84E-03
Brucite-3.0-1 60C	60	3.2	105	9.03	9.43	2.72E-03
			127	9.09	9.49	2.68E-03
			342	9.14	9.54	3.01E-03
Brucite-3.0-2 60C	60	3.2	105	9.02	9.43	2.72E-03
			127	9.09	9.49	2.69E-03
			342	9.11	9.51	2.85E-03
Brucite-4.0-1 60C	60	4.4	105	8.93	9.52	2.70E-03
			127	9.01	9.60	2.53E-03
			342	9.06	9.65	2.82E-03
Brucite-4.0-2 60C	60	4.4	105	8.86	9.45	2.63E-03
			127	8.99	9.58	2.56E-03
			342	9.07	9.66	2.64E-03
Brucite-5.0-1 60C	60	5.6	105	8.78	9.48	2.44E-03
			127	8.85	9.55	2.39E-03
			342	8.95	9.65	2.60E-03
Brucite-5.0-2 60C	60	5.6	105	8.80	9.50	2.36E-03
			127	8.86	9.56	2.39E-03
			342	8.95	9.65	2.47E-03

Table 3. (Continued)

Exp. #	T, C	NaCl, mol•kg^{-1}	Exp. time, d	pH	pH$_m$	$m_{\Sigma Mg}$, mol•kg^{-1}
Brucite-0.01-1 70C	70	0.01	104	8.97	8.97	1.68E-03
			126	9.02	9.02	1.72E-03
			342	9.02	9.02	1.84E-03
Brucite-0.01-2 70C	70	0.01	104	8.95	8.95	1.65E-03
			126	9.04	9.04	1.64E-03
			342	9.02	9.02	1.71E-03
Brucite-0.1-1 70C	70	0.1	104	9.00	9.00	1.87E-03
			126	9.08	9.08	1.83E-03
			342	9.06	9.06	2.00E-03
Brucite-0.1-2 70C	70	0.1	104	9.07	9.07	2.15E-03
			126	9.09	9.09	1.94E-03
			342	9.07	9.07	2.09E-03
Brucite-1.0-1 70C	70	1.0	104	8.95	9.05	2.58E-03
			126	9.01	9.11	2.45E-03
			342	9.04	9.14	2.76E-03
Brucite-1.0-2 70C	70	1.0	104	8.93	9.03	2.60E-03
			126	9.05	9.15	2.56E-03
			342	9.03	9.13	2.71E-03
Brucite-2.0-1 70C	70	2.1	104	8.83	9.04	2.64E-03
			126	8.98	9.19	2.54E-03
			342	8.98	9.19	2.80E-03
Brucite-2.0-2 70C	70	2.1	104	8.86	9.07	2.67E-03
			126	8.98	9.19	2.57E-03
			342	8.98	9.19	2.87E-03
Brucite-3.0-1 70C	70	3.2	104	8.74	9.13	2.60E-03
			126	8.89	9.28	2.64E-03
			342	8.82	9.21	2.73E-03
Brucite-3.0-2 70C	70	3.2	104	8.74	9.13	2.62E-03
			126	8.89	9.28	2.70E-03
			342	8.82	9.21	2.62E-03

Exp. #	T, C	NaCl, mol•kg^{-1}	Exp. time, d	pH	pH$_m$	$m_{\Sigma Mg}$, mol•kg^{-1}
Brucite-4.0-1 70C	70	4.4	104	8.63	9.18	2.47E-03
			126	8.85	9.40	2.43E-03
			342	8.76	9.31	2.67E-03
Brucite-4.0-2 70C	70	4.4	104	8.63	9.18	2.49E-03
			126	8.84	9.39	2.57E-03
			342	8.77	9.32	2.60E-03
Brucite-5.0-1 70C	70	5.6	104	8.63	9.34	2.38E-03
			126	8.76	9.47	2.33E-03
			342	8.65	9.38	2.54E-03
Brucite-5.0-2 70C	70	5.6	104	8.59	9.30	2.50E-03
			126	8.72	9.43	2.51E-03
			342	8.64	9.35	2.44E-03
Brucite-0.01-1 80C	80	0.01	72	8.84	8.84	1.51E-03
			120	8.84	8.84	1.58E-03
Brucite-0.01-2 80C	80	0.01	72	8.85	8.85	1.60E-03
			120	8.92	9.02	1.74E-03
Brucite-0.1-1 80C	80	0.1	72	8.89	8.89	1.79E-03
			120	8.81	8.81	1.93E-03
Brucite-0.1-2 80C	80	0.1	72	8.88	8.88	1.76E-03
			120	8.98	8.98	2.02E-03
Brucite-1.0-1 80C	80	1.0	72	8.80	8.90	2.30E-03
			120	8.84	8.94	2.81E-03
Brucite-1.0-2 80C	80	1.0	72	8.79	8.89	2.32E-03
			120	8.90	9.00	2.64E-03
Brucite-2.0-1 80C	80	2.1	72	8.85	9.07	2.57E-03
			120	8.84	9.06	2.93E-03
Brucite-2.0-2 80C	80	2.1	72	8.86	9.08	2.61E-03
			120	8.83	9.05	3.26E-03
Brucite-3.0-1 80C	80	3.2	72	8.72	9.14	2.37E-03
			120	8.70	9.12	2.61E-03
Brucite-3.0-2 80C	80	3.2	72	8.71	9.13	2.40E-03
			120	8.79	9.21	2.75E-03
Brucite-4.0-1 80C	80	4.4	72	8.60	9.16	2.32E-03
			120	8.69	9.25	2.78E-03

Table 3. (Continued)

Exp. #	T, C	NaCl, mol•kg^{-1}	Exp. time, d	pH	pH_m	$m_{\Sigma Mg}$, mol•kg^{-1}
Brucite-4.0-2 80C	80	4.4	72	8.61	9.17	2.32E-03
			120	8.71	9.27	2.63E-03
Brucite-5.0-1 80C	80	5.6	72	8.52	9.17	2.15E-03
			120	8.65	9.30	2.41E-03
Brucite-5.0-2 80C	80	5.6	72	8.50	9.17	2.13E-03
			120	8.64	9.29	2.46E-03

The equilibrium quotients for Reaction (3) determined in this study are tabulated in Table 4.

Table 4. Equilibrium quotients for Reaction (3) determined in this study

T, °C	m_{NaCl}, mol•kg^{-1}	log Q_{s2}, ± 2σ
40	0.010	16.57 ± 0.09
	0.10	16.85 ± 0.09
	1.0	16.92 ± 0.15
	2.1	17.07 ± 0.13
	3.2	17.03 ± 0.25
	4.4	17.15 ± 0.27
	5.6	17.23 ± 0.22
50	0.010	16.17 ± 0.16
	0.10	16.46 ± 0.33
	1.0	16.42 ± 0.17
	2.1	16.63 ± 0.30
	3.2	16.51 ± 0.61
	4.4	16.64 ± 0.32
	5.6	16.74 ± 0.38
60	0.010	15.74 ± 0.18
	0.10	15.88 ± 0.18
	1.0	16.11 ± 0.25

T, °C	m_{NaCl}, mol•kg^{-1}	log Q_{s2}, ± 2σ
	2.1	16.17 ± 0.27
	3.2	16.37 ± 0.21
	4.4	16.56 ± 0.32
	5.6	16.51 ± 0.31
70	0.010	15.21 ± 0.15
	0.10	15.39 ± 0.15
	1.0	15.60 ± 0.20
	2.1	15.70 ± 0.28
	3.2	15.81 ± 0.27
	4.4	15.97 ± 0.38
	5.6	16.13 ± 0.24
80	0.010	14.89 ± 0.20
	0.10	15.02 ± 0.29
	1.0	15.24 ± 0.26
	2.1	15.54 ± 0.06
	3.2	15.64 ± 0.20
	4.4	15.79 ± 0.29
	5.6	15.79 ± 0.37

Using the SIT coefficients, $\varepsilon(Mg^{2+}, Cl^-)$ and $\varepsilon(H^+, Cl^-)$, from Xiong (2006), at the experimental temperatures for this work, the equilibrium quotients in Table 4 are extrapolated to infinite dilution based on the following equation,

$$\log_{10} K^0_{s2} = \log_{10} Q_{s2} - 2D + \varepsilon(Mg^{2+}, Cl^-) I_m - 2\varepsilon(H^+, Cl^-) I_m + 2\log_{10} a_{H_2O} \quad (8)$$

where $\log K^0_{s2}$ is an equilibrium constant at infinite dilution for Reaction (3); $\log Q_{s2}$ an equilibrium quotient for Reaction (3) at certain ionic strengths; D the Debye-Hückel term defined by the following equation; I_m ionic strength on a molal scale; a_{H_2O} activity of water,

$$D = \frac{A_\gamma \sqrt{I_m}}{1 + 1.5 \times \sqrt{I_m}} \tag{9}$$

where A_γ is Debye-Hückel slope for activity coefficients.

In using Equations (8) and (9) to extrapolate to infinite dilution, the activities of water in NaCl solutions under the experimental conditions are from Gibbard et al., (1974), and the Debye-Hückel slopes for activity coefficients at the experimental temperatures of this work are from Helgeson et al., (1981). The equilibrium constants for Reaction (3) at infinite dilution are listed in Table 5.

In Figure 4, the solubility constants at infinite dilution as a function of reciprocal temperature are displayed. In this figure, the $\log_{10} K^0_{s2}$ at 25°C from Xiong (2008) is also included. The temperature function of $\log_{10} K^0_{s2}$ can be described by the following expression:

$$\log_{10} K^0_{s2} = \frac{4306.8369}{T} + 2.6383 \tag{10}$$

where T is temperature in K. In the linear regression, the $\log_{10} K^0_{s2}$ at 25°C from Xiong (2008) is also incorporated.

Table 5. Equilibrium constants for Reaction (3) at infinite dilution determined in this work

Temperature, °C	$\log_{10} K^0_{s2} \pm 2\sigma$
40	16.43 ± 0.22
50	15.96 ± 0.32
60	15.63 ± 0.11
70	15.12 ± 0.08
80	14.84 ± 0.16

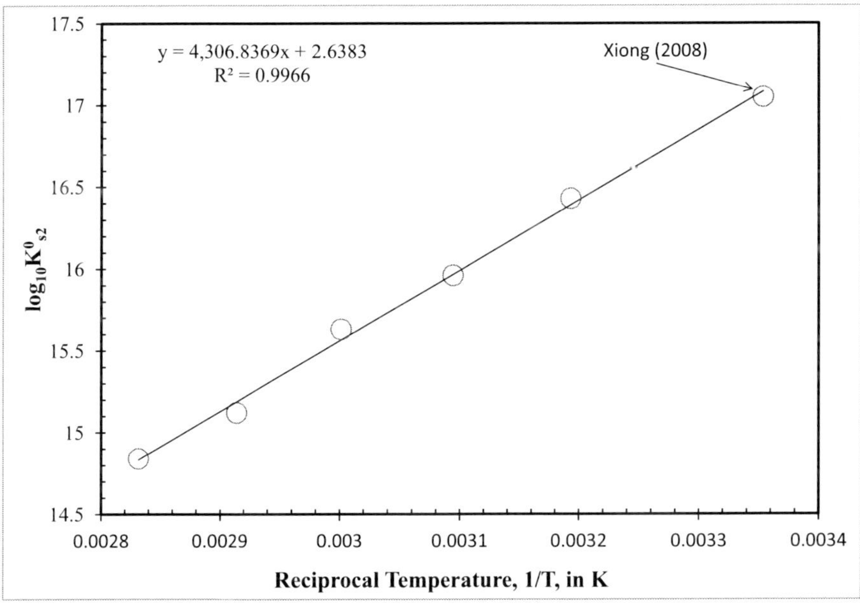

Figure 4. A plot showing the solubility constants at infinite dilution as a function of reciprocal temperature. Notice that the data point at 25°C from Xiong (2008) is also included in the regression.

CONCLUSION

In this study, we have determined the solubility constants of brucite in NaCl solution at elevated temperatures up to 80°C. The hydrogen ion concentrations in our experiments in NaCl solutions to high ionic strength at elevated temperatures are determined by applying the correction factors that are experimentally acquired in this work. As brines associated with salt formations are of high ionic strength in nature, the results from our work are applicable to nuclear waste isolation in salt formations. They are also relevant for modeling the formation of brucite-carbonate chimneys in hyperalkaline (pH 9-11), saline, hydrothermal fluids (40°C to 90°C). Based on the highly linear relation between the solubility constant of brucite and temperature, the solubility constants of brucite can be accurately extrapolated down to 0°C, and up to 100°C.

ACKNOWLEDGMENTS

Sandia National Laboratories is a multi-mission laboratory managed and operated by National Technology and Engineering Solutions of Sandia, L.L.C., a wholly owned subsidiary of Honeywell International, Inc., for the U.S. Department of Energy's National Nuclear Security Administration under contract DE-NA-0003525. This research is funded by the Salt R&D programs administered by the Office of Nuclear Energy (NE) of the U.S. Department of Energy.

REFERENCES

Altmaier, M., V. Metz, V. Neck, R. Müller, and Th Fanghänel, 2003. "Solid-liquid equilibria of $Mg(OH)_2(cr)$ and $Mg_2(OH)_3Cl{\bullet}4H_2O(cr)$ in the system Mg-Na-H-OH-Cl-H_2O at 25°C." *Geochimica et Cosmochimica Acta* 67, 3595–3601.

Behlau, J. and Mingerzahn, G., 2001. "Geological and tectonic investigations in the former Morsleben salt mine (Germany) as a basis for the safety assessment of a radioactive waste repository." *Engineering Geology* 61, 83–97.

Brewitz, W., Droste, J. and Stier-Friedland, G., 2008. "Geological features of the Morsleben repository and their relevance for long-term safety." *Reviews in Engineering Geology* 19, 53-66.

Gibbard Jr, H. Frank, George Scatchard, Raymond A. Rousseau, and Jefferson L. Creek, 1974. "Liquid-vapor equilibrium of aqueous sodium chloride, from 298 to 373. deg. K and from 1 to 6 mol kg^{-1}, and related properties." *Journal of Chemical and Engineering Data* 19, 281–288.

Helgeson, Harold C., David H. Kirkham, and George C. Flowers, 1981. "Theoretical prediction of the thermodynamic behavior of aqueous electrolytes by high pressures and temperatures; IV, Calculation of activity coefficients, osmotic coefficients, and apparent molal and standard and relative partial molal properties to 600 degrees C and 5kb." *American Journal of Science* 281, 1249–1516.

Krumhansl, J. L., Panenguth, H. W., Zhang, P. C., Kelly, J. W., Anderson, H. L., Hardesty, J. O., 2000. "Behavior of MgO as a CO_2 scavenger at the Waste Isolation Pilot Plant (WIPP), Carlsbad, New Mexico." *Materials Research Society Symposium Proceedings* 608, 155–160.

Monnin, C., V. Chavagnac, C. Boulart, B. Ménez, Martine Gérard, E. Gérard, C. Pisapia, M. Quemeneur, G. Erauso, A. Postec, and L. Guentas-Dombrowski, 2014. "Fluid chemistry of the low temperature hyperalkaline hydrothermal system of Prony Bay (New Caledonia)." *Biogeosciences* 11, no. 20, 5687.

Okumura, T., Y. Ohara, R. J. Stern, T. Yamanaka, Y. Onishi, H. Watanabe, C. Chen, S. H. Bloomer, I. Pujana, S. Sakai, T. Ishii, 2016. "Brucite chimney formation and carbonate alteration at the Shinkai Seep Field, a serpentinite-hosted vent system in the southern Mariana forearc." *Geochemistry, Geophysics, Geosystems* 17, 3775–3796.

Palmer, Donald A., and David J. Wesolowski, 1997. "Potentiometric measurements of the first hydrolysis quotient of magnesium (II) to 250 C and 5 molal ionic strength (NaCl)." *Journal of Solution Chemistry* 26, 217–232.

Schuessler, W., B. Kienzler, S. Wilhelm, V. Neck, and J. I. Kim, 2001. "Modeling of Near Field Actinide Concentrations in Radioactive Waste Repositories in Salt Formations: Effect of Buffer Materials." *Mat. Res. Soc. Symp. Proc.* 663, 791.

Schwartz, M. O., 2012. Modelling groundwater contamination above a nuclear waste repository at Gorleben, Germany. *Hydrogeology Journal* 20, 533-546.

Sőhnel, O, Novotný, P., 1985. *Densities of aqueous solutions of inorganic substances*. Elsevier, New York, 335 p.

Stewart, S., 2002. Exploring the continental shelf for low geological risk nuclear waste repository sites using petroleum industry databases: a UK case study. *Engineering Geology* 67, 139-168.

U.S. National Academy of Sciences Committee on Waste Disposal. 1957. *The Disposal of Radioactive Waste on Land*. Publication 519.

Washington, DC: National Academy of Sciences–National Research Council.

Wood, S. A., Palmer, D. A., Wesolowski, D. J., Bénézeth, P., 2002. "The aqueous geochemistry of the rare earth elements and yttrium. Part XI. The solubility of $Nd(OH)_3$ and hydrolysis of Nd^{3+} from 30 to 290°C at saturated water vapor pressure with in-situ pHm measurement." In: Hellmann, R. and Wood, S. A., (Eds.), *Water-Rock Interactions, Ore Deposits, and Environmental Geochemistry: A Tribute to David Crerar,* Special Publication 7, The Geochemical Society, pp. 229–256.

Xiong, Y. L., 2003. "Revised thermodynamic properties of brucite determined by solubility studies and its significance to nuclear waste isolation." In *Abstracts with Programs, Geological Society of America Annual Meeting (Seattle, Washington*, vol. 35, no. 6, pp. 102-103.

Xiong, Y. L., 2006. "Estimation of medium effects on equilibrium constants in moderate and high ionic strength solutions at elevated temperatures by using specific interaction theory (SIT): interaction coefficients involving Cl^-, OH^- and Ac^- up to 200°C and 400 bars." *Geochemical Transactions* 7, no. 1: 4.

Xiong, Yongliang, 2008. "Thermodynamic properties of brucite determined by solubility studies and their significance to nuclear waste isolation." *Aquatic Geochemistry* 14, 223–238.

Xiong, Y. L., and A. C. S. Lord, 2008. "Experimental investigations of the reaction path in the $MgO–CO_2–H_2O$ system in solutions with ionic strengths, and their applications to nuclear waste isolation." *Applied Geochemistry* 23, 1634–1659.

Xiong, Y. L., H. Deng, M. Nemer, S. Johnsen, 2010. "Experimental determination of the solubility constant for magnesium chloride hydroxide hydrate ($Mg_3Cl(OH)_5•4H_2O$, phase 5) at room temperature, and its importance to nuclear waste isolation in geological repositories in salt formations." *Geochimica et Cosmochimica Acta* 74, 4605–4611.

In: Solution Chemistry
Editor: Yongliang Xiong
ISBN: 978-1-53613-101-7

Chapter 3

CONFINED WATER IN ROCK SALT

Mei Ding[1], Bernard H. W. S. de Jong[2], Erik Watkins[1], Monika Hartl[3], Luc Daemen[4], Hylke. N. de Jong[5] and Qingcheng He[6]

[1]Los Alamos National Laboratory, Los Alamos, NM, US
[2]Institute of Earth Sciences, Utrecht University, the Netherlands
[3]European Spallation Source, Lund, Sweden
[4]Oak Ridge National Laboratory, Oak Ridge, TN, US
[5]Dept. of Anthropology, Rutgers University, New Brunswick, NJ, US
[6]Chinese Academy of Geological Sciences, Beijing, China

ABSTRACT

Moisture on grain boundaries of salt crystals is the principal expedient in halokinesis, the rheology of rock salt, yet its quantitative determination has proven to be elusive. Here we elaborate on results from a first-of-its-kind incoherent inelastic neutron scattering (INNS) of bedded salts to distinguish three water environments: intergranular water molecules confined to grain boundaries, water trapped as brine in fluid inclusions, and structural water in hydrous minerals (Ding et al., 2015). Sixteen spectral lines can be distinguished unambiguously in the 0-1100 cm^{-1} multiphonon and librational domain, yielding an unprecedented high resolution for a natural material. The spectral response to temperature illustrates the bimodality of the technique enabling the intergranular water component to be distinguished from that of the brine, shedding light on a nearly 30-year old problem in characterizing different forms of water in rock salt. This pioneering study shows that INNS provides insight into the cause and effect of moisture migration and its coupling to thermomechanical properties in salt formation. Our results are pertinent to subsurface energy exploration and storage, including nuclear waste storage in salts.

INTRODUCTION

Since the moment that organisms left the seas they evolved in, the acquisition of salt has been a principal preoccupation. Salt's discrete distribution and it's nutritional (and later industrial) importance has led to extensive study of the geology of its occurrence in salt lakes, playas (desert flats), sabkhas (coastal salt flats) and subsurface deposits. Salt deposits form commonly in continental tectonic basins, wind deflation hollows, and abandoned fluvial valleys in the world's desert belt. The influence of salt on animal physiology, sociology, kindship relations, and migration patterns is a burgeoning field (Baas Becking 1931, Haak et al. 2008, de Jong 2017, Bloch 1963, Denton 1982, Multhauf 1978) and its historic and prehistoric importance is well documented (Flad 2011, Harding 2013, Multhauf 1978, Nenquin 1961).

In arid climates, evaporation of water bodies is the sole process by which salt is formed. The primary scenario for rock salt formation is a barrier reef with an inland sea: infill only happens at times of complete basin isolation, there is no surface connection to the world's oceans (Ochsenius 1877, Ochsenius 1906) and where solubility products are superseded. Similarly, salt genesis either manmade or natural occurs in salt pans where brine is evaporated, often paired with a characteristic algal coloration ranging from yellow to dark red (Baas Becking 2015).

Tamman (Tamman 1903) recognized that precipitation involves two steps, a nucleation step indicative of the degree of oversaturation and a growth step usually at higher temperature at which the nutrients diffuse towards the nucleus. The fewer the number of nuclei, the smaller the degree of oversaturation, the coarser the resulting precipitate. In the course of this process supernatant water is occluded in the commonly dendritic crystallites formed at the air-saline water interface and results in fluid inclusions in the final precipitate upon gravitational settling (Llewellyn 1968, Aquilano, et al., 2016, Leitner, et al., 2013, Rahimpour-Bonab et al., 2007). The abundance of fluid inclusions and the variation in crystal morphology and texture are indicative of the specific brine conditions under which the crystallites are formed. These inclusions are mobile and may migrate to grain boundaries due to differential stress or temperature gradients.

The precipitation sequences have been thoroughly studied. From a 1000 meter seawater column about 7 cm carbonate precipitates first, followed by 60 cm gypsum, 12 meter of halite and 3 meters of K/Mg salts as chlorides and sulfates. To account for the observed bed thicknesses of kilometers, eustatic sea level rises must have occurred to replenish the basin, starting the precipitation cycle anew. It is estimated that a 1km thick Zechstein evaporate deposit takes about 300.000 years to form (Börner, 2003). In the late Miocene Messinian crisis in the Mediterranean 2 km thick halite-gypsum deposits were deposited in 300.000 years.

Hugh rock salt deposits are by no means monomineralic, besides replenishing the basin causing a new precipitation cycle, siliciclastic runoff from the surrounding basin rim, blown in quartz, feldspar, clays, or zeolites, and volcanic ash result in an insoluble residue.

Seven major rock salt basins occur in North America: the Carboniferous Sverdrup basin in Northern Canada far above the arctic circle North of 74 degrees latitude; the Devonian Elk Point in Sasketchwan; the Devonian Jurassic Williston in Sasketchwan, Montana and the Dakotas; the Pennsylvania Paradox basin in Utah; the Permian Delaware basin in Texas and New Mexico; the Jurassic Gulf Coast basin; and the Silurian-Devonian Appalachian Salina basin. During the 1950s these salt deposits, either bedded domal ordiapiritic, were considered good candidates as possible nuclear waste disposal sites, because of the extreme low permeability (10^{-21} m^2) of intact rock salt which imparts excellent sealing capacity for instance as cap rocks in hydrocarbon reservoirs. Also, salt mines tend to be very dry places (Urai and Spiers 2007, Roedder and Basset 1981).

It was recognized early on that the role of this occluded moisture, intergranular or within fluid inclusions was the principal unknown in delineating natural halokinesis in the presence of thermal gradients induced by nuclear waste canisters or changes in overburden load by ice sheet removal. Besides taking into account formation heterogeneity with concomitant blockage of crack tips, the formation and determination of the amount of intergranular moisture became the central focus and proved difficult to determine as already pointed out by Roedder and Basset (Roedder and Basset 1981). Every conceivable technique, infrared, Raman, proton NMR in combination with developments in microstructural and textural/orientation analysis such as electron backscatter diffraction, microstructure enhancement by gamma radiation or cryo scanning electron microscope (SEM) have been brought to bear on its determination. Despite these efforts grain boundary moisture and its effect on halikonesis continues to be an issue.

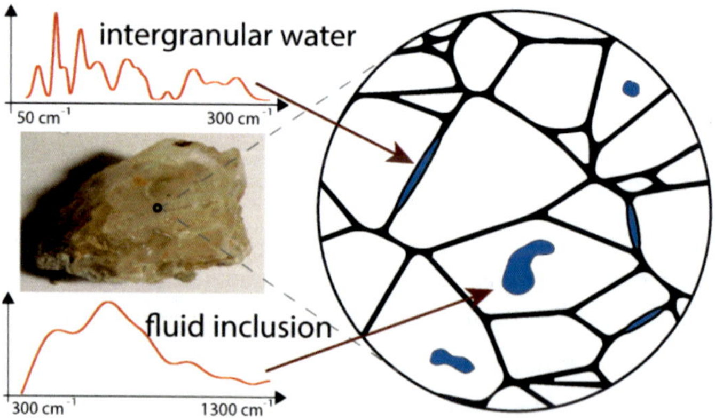

Figure 1. Incoherent inelastic neutron spectrum of rocks salt showing the spectral discrimination between intergranular and fluid inclusion moisture. The mineral fabric is a slice through a 3D Voronoi simulation generated with random nucleation sites.

Here we have addressed this issue employing a novel technique: inelastic incoherent neutron scattering (IINS). Figure 1 shows that with this technique it is possible to unequivocally separate the signal due to intergranular moisture from that in the fluid inclusions. The provenance of the samples is from the Zechstein Delaware basin in New Mexico, specifically the Permian bedded salt at the Salado formation. As such it is a unique experiment making it possible for the first time to unequivocally determine quantitatively moisture in grain boundaries.

APPLICATION OF NEUTRON TECHNOLOGIES IN CHARACTERIZING WATER IN A CONFINED ENVIRONMENT

Why Neutrons?

A wide spectrum of analytical techniques is used to characterize and determine water in rock salt, including (but not limited to) gravimetry (Roedder and Basset 1981), microscopy (Olander et al. 1981), infrared (IR) (Fayer and Levinger 2010), Raman, NMR (Baumgartner and Bakker 2010), synchrotron-based X-rays (Maccarini 2007), Laser Ablation ICP-MS (Sun et al., 2013), and micrography (Lowenstein et al., 2001, Timofeeff et al., 2001). Each of these techniques illuminates a unique probe into the characteristics and behavior of water in rock salt.

On top of these, neutron and X-ray scattering are the two principal techniques capable of providing atomic scale properties of fluids and their reactivity (Willis and Carlile 2009, Als-Nielsen and McMorrow 2000). The drawback of X-ray scattering is its insensitivity to light elements. X-ray scattering intensity scales roughly with the atomic number of an element (Figure 2). Neutrons on the other hand are scattered by atomic nuclei. Their scattering intensity is a non-linear function of atomic mass, making neutrons particularly sensitive to elements with large cross-sections, specifically hydrogen (Sears 1992). The distinct advantage in using neutrons to characterize fluids (e.g., H_2O) has been recognized as the unique advanced tool for characterizing confined moisture in structures and at interfaces (Cole et al., 2006).

Neutron Techniques: Neutron Diffraction, Small-Angle Scattering, and Inelastic Spectroscopy

Among neutron techniques, coherent elastic neutron diffraction including small-angle neutron scattering (SANS) is the most important tool for confined water characterization. The physics behind itis laid down in Bragg's law ($n\lambda = 2d\sin\theta$) (Cowley 1975, Willis and Carlile 2009) as illustrated in Figure 3. It allows assessment of confined liquids and probing of structural correlations at the molecular scale. For SANS measurements, the beam samples continuously greater distances when approaching the beam target. These increasing distances are not related to d-spacing in crystalline structure, i.e., the scattering is not from individual atoms or aggregates of atoms. The scattering takes place at a coarser level, assemblies of particles, density fluctuations in solids or fluids corresponding to a large d (pore size) in real space, allowing study of both structure and form factors.

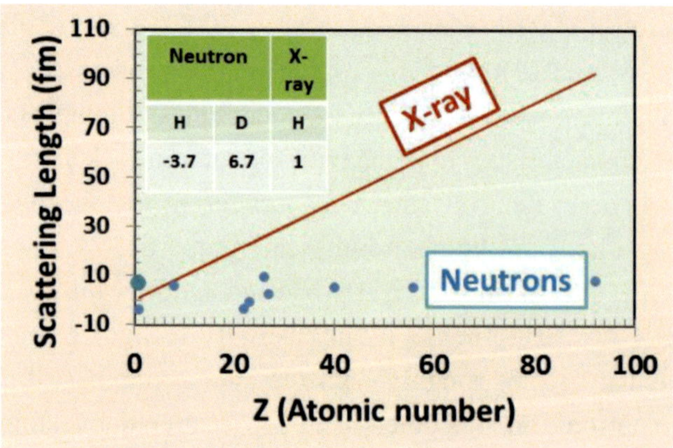

Figure 2. X-ray and neutron atomic scattering length. Note the anomalous values for Hydrogen (Als-Nielsen and McMorrow 2000, Sears 1992).

Neutron diffraction has been applied primarily in studies of confined liquids involving water-ice transitions in porous media because of its unique capability to assess structural characteristics of supercooled confined water (Dore 2000).

SANS in combining with ultra-SANS(USANS) techniques have been recognized in the last ten years as the canonical technique in pore characterization from nanometers to tenths of micrometers (Radlinski et al., 1996, Anovitz et al., 2009), to study water accessibility in various confining environments, such as water in clay (Ding et al., 2013), natural rock and particularly shale reservoirs (Gu et al., 2016). This technique is a unique and particularly advanced tool in the study of pore morphology controlled fluids transport and interaction in solids.

Incoherent inelastic neutron scattering (IINS), a vibrational spectroscopic method, is an energy loss spectroscopy in which a fraction of the incident neutrons loses energy by exciting vibrational modes of the scatterer (Mitchel et al., 2003, Willis and Carlile 2009, Tarjus et al., 1993). It is selectively sensitive to the dynamics of hydrogen atoms, has no selection rules, and is nondestructive and thus easily penetrates large samples (Pechar et al., 1982, Li et al., 1991). These attributes makes its use advantageous relative to Infrared and Raman spectroscopy which measure vibrations indirectly through oscillations of the dipole moment vector or of the polarization tensor of the liquid sample, lack the penetration depth and are subjected to selection rules. The neutron sensitivity to the dynamics of hydrogen atoms is the consequence of their extremely high incoherent scattering cross-section which is about 80 times than that of other typical elements in compounds. As such, the librational modes of water, which are either weak or inactive in optical vibrational spectroscopy (e.g., Raman or FTIR) are readily observed with neutrons. IINS is a powerful technique and has been applied successfully to investigate the signature of various forms of water in geological materials (Mitchell et al., 2003, Ockwig et al., 2008, Ding et al., 2015). Furthermore, due to its high penetration and non-destructive nature, neutron based techniques allow testing much larger intact samples at elevated temperature and pressure. It is therefore particularly suited for studying geologic material relevant to nuclear waste repositories where heterogeneity, elevated temperatures, and pressurized conditions prevail. The work we present in this paper is the first exploratory effort using neutron scattering to characterize water in rock salt.

Inelastic Spectroscopy Study of Water in Permian Salt at High Temperature

Background

The Permian Zechstein salts of the Salado formation in Southern New Mexico, US, are the site of the Waste Isolation Pilot Plant (WIPP) (http://www.wipp.energy.gov), the only licensed deep geologic repository for permanent disposal of defense transuranic (TRC) waste in the USA. This, and similar, rock salt formations are currently under consideration for storage of Heat-Generating Nuclear Waste (HGNW) (Hansen and Leigh 2011, Robinson et al., 2012, Wald 2014), compelling a re-assessment of its rheology at higher temperatures (Carter and Hansen 1983). Thus, the presence of water in bedded Permian salts and its behavior upon heating determines the suitability as a nuclear waste storage medium for the disposal of HGNW.

Water in salt and its role in deformation and weakening of rock salt during long-term creep have been studied extensively (Carter and Hansen 1983, Schoenherr et al., 2007, Muhammed 2015, Knauth and Kumer 1981, Roedder 1984, Urai 1983, Urai et al., 1986, Spiers et al., 1988, Peach and Spiers 1995, Houben et al., 2013). In addition, brine migration in salt under thermal gradients have been reported by numerous researchers (Jordan et al., 2015, Bradshaw and Sanchez 1969, Roedder and Belkin 1980, Cheung et al., 1980, Pigford 1982, Hwang et al., 1990).

Previous studies indicate that water occurs in rock salt in three forms: bound water in hydrous minerals (e.g., clays, hydrated salts) (Braitsch 1971, Roedder and Bassett 1981, Jordan et al., 2015, Roedder 1984), intergranular pore water in grain boundaries (Roedder 1984, Desbois, et al., 2012) and free water trapped as brine in fluid inclusions (Roedder 1984, Pigford 1982, Hwang et al., 1990). However, as pointed out three decades ago by Roedder and Bassett (Roedder and Bassett 1981) errors and problems persist in assigning these three water forms unambiguously leading to incompatibilities among published results.

Two major problems impede progress in determining water content and study of its transport behavior in rock salt. One is the experimental lack of a technique to distinguish and quantify the three different water forms, particularly that for intergranular water. The second problem is intrusive sample preparation required by analytical techniques, causing changes in brine analysis (Roedder and Bassett 1981). However, progress in micrography has made chemical analysis of 30 micron brine inclusions feasible (Lowenstein et al., 2001, Timofeeff et al., 2001). Another important unexplored issue is the effect of water release and movement of nanoscale pore water confined in grain boundaries as opposed to bulk water, on the coupled thermal-mechanical behavior of intact and crushed salt (Fayer and Levinger 2010).

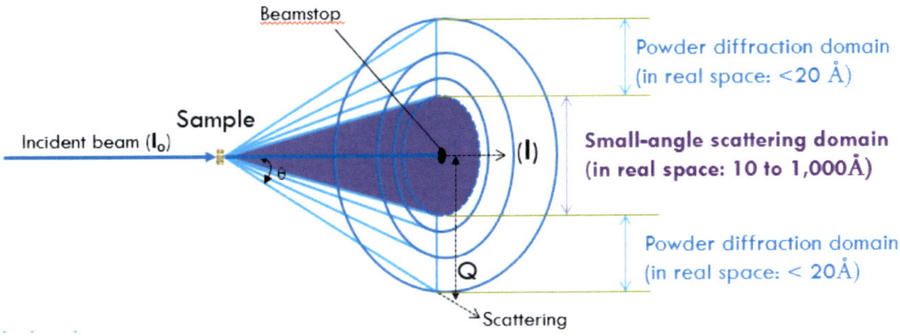

Figure 3. Illustration of elastic experimental diffraction. The scattering is measured in the Q plane. The distances in this plane are inversely related to those of the sample and span the reciprocal space, from which distances in real space are calculated.

Figure 4. Image of fluid inclusion in Permian salt.

The total water content in bedded Permian salt various between samples from 0.05 to 0.5 wt.% (Caporuscio et al., 2014). Brine present in the form of fluid inclusions from few microns to several millimeters (Figure 4) and its migration under temperature gradients has been studied over and over again (Roedder and Belkin 1979, 1980, Cheung et al., 1980, Hwang et al., 1990, Caporuscio et al., 2014, Jordan 2015). Bound water in hydrous minerals present in gypsum, polyhalite, and as interlayer water in clays has also been investigated repeatedly. However, characterization of intergranular nanoscale pore water confined in grain boundaries and its release and migration behavior under high temperature condition has turned out to be not feasible.

To remedy this knowledge gap we have carried out Incoherent Inelastic Neutron Scattering (IINS), using the Filter Difference Spectrometer (FDS) at the Los Alamos Neutron Science Center (LANSCE). Our goal was to characterize specific water forms in rock salt and investigate the release of water including nanoscale pore confined water at elevated temperature. We use this technique in a first exploratory effort to characterize water in rock salt. Our principal result shows that the observed spectral lines discriminate unambiguously between multiphonon, librational, and O-H stretching modes of intergranular pore water, brines, and bound water in hydrous minerals. This capability has profound implications in unveiling the fundamental aspects of moisture migration in rock salt; a prerequisite for safe and robust sequestration of HGNW in bedded salt formations.

Materials and Methods

Two samples were used in this study. One is unadulterated rock salt from the Permian salt formations located near Carlsbad, New Mexico, the other its insoluble residue (Figure 5a and 5b). No specific sample preparation requirements for IINS measurement are needed as long as it fits the instrument cylinder sample cell (ϕ2 cm, L15 cm), which can hold up to 150 gram intact pristine or processed rock salt. Such simple and nondestructive sample preparation set up avoids loss of intergranular pore water in grain boundaries and fluid inclusions during preparation. It also minimizes the impact of field sample heterogeneity, significantly improving and advancing knowledge concerning the water content of rock salt samples for nuclear waste storage. However, the sample must be carefully selected as it is well known since the pioneering studies of Usiglio (Usiglio 1849) and van‘t Hoff (Hoff 1905) that salt deposits are laterally inhomogeneous; calcium-magnesium carbonates as least soluble minerals precipitating first, followed by gypsum, halite, magnesium sulfate, potassium chloride, and magnesium chloride, as most soluble, precipitating last in the center of the basin. These laboratory determined sequences were welcomed by Ochsenius (Ochsenius 1906) as confirmations of his observations in nature, specifically the Kara Bogaz Gol, a shallow inundated depression halfway the Eastern shore rim of the Caspian sea.

The Permian bedded salt in the WIPP area is inhomogeneous, mainly halite (NaCl) and polyhalite ($K_2Ca_2Mg(SO_4)_4 \cdot 2(H_2O)$, with minor amounts of magnesite ($MgCO_3$), clays, gypsum ($CaSO_4.2H_2O$), and quartz (SiO_2) (Caporuscio et al., 2014). This phase assemblage is characteristic of late Precambrian, Permian, and ternary seawater precipitates with high Mg/Ca ratios, polyhalite being one of the characteristic phases (Lowenstein et al., 2001). The residual (~0.5 weight%) consists primarily of quartz and corrensite (an interlayered smectite/chlorite swelling clay) and kaolinite, with still some halite and magnesite.

We anticipated that intergranular pore water and fluid inclusions would be removed during dissolution. In this manner we can isolate chemically and physically bound water present in hydrous minerals from the original rock salt.

Dissolution and Separation of the Rock Salt

Past studies have shown that Permian salt in the WIPP area consists mainly of halite (NaCl) and polyhalite ($K_2Ca_2Mg(SO_4)_4 \cdot 2(H_2O)$, with minor amounts of magnesite ($MgCO_3$), clays (e.g., mica, smectite, kaolinite), gypsum ($CaSO_4.2H_2O$), and quartz (SiO_2) (Caporuscio et al.,

2014). The large solubility difference between these minerals enables separating the insoluble phases including hydrous minerals such as clays from the original rock salt. In this study, about 500 gram of rock salt (Figure 5a) was dissolved in 1 liter distilled water, stirred gently until all coarse salt sample was dissolved. The suspension was filtered using 0.45 μm filter paper. The residual was collected on the filter paper and vacuum dried at 110°C overnight (Figure 5b).

Thermo Gravimetric Analysis (TGA)

The mass loss and decomposition of rock salt and its insoluble residue as a function of temperature were determined using a Netzsch Jupiter STA-449 Thermal Analyzer. The weight loss curves were obtained by placing 4 mg samples in an alumina sample holder, with a heating rate of 10°C/min, under nitrogen gas flow.

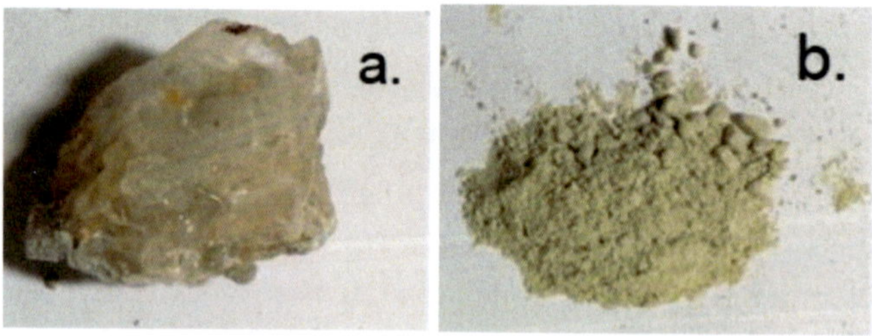

Figure 5. Image of (a) Permian rock salt and (b) its insoluble residue.

IINS Spectroscopy

The Filter Difference Spectrometer (FDS) at LANSCE of Los Alamos National Laboratory (LANL) is used for IINS molecular vibrational spectroscopy. The instrument (Figure 6) is designed to provide high count rates using a large solid angle detector. Vibrational spectra are obtained by measuring neutron energy loss. The FDS determines the energy change by the use of low energy band-pass filters (Be) between sample and detector. The scattered neutrons are sorted by their total time of flight as they journey from source to detector, from which the incident energies can be determined. Data reduction involves variable numerical deconvolution of the spectrometer response function. Technical details of FDS can be found at http://lansce.lanl.gov/lujan/instruments/FDS.shtml. The samples of room temperature (RT) and those heat treated at 100° and 150°C for rock salt, at 150°C and 450°C for the insoluble residue, were loaded in cylindrical aluminum sample holders, mounted in a closed-cycle refrigerator, and measured at 10K. A vanadium rod was used as background.

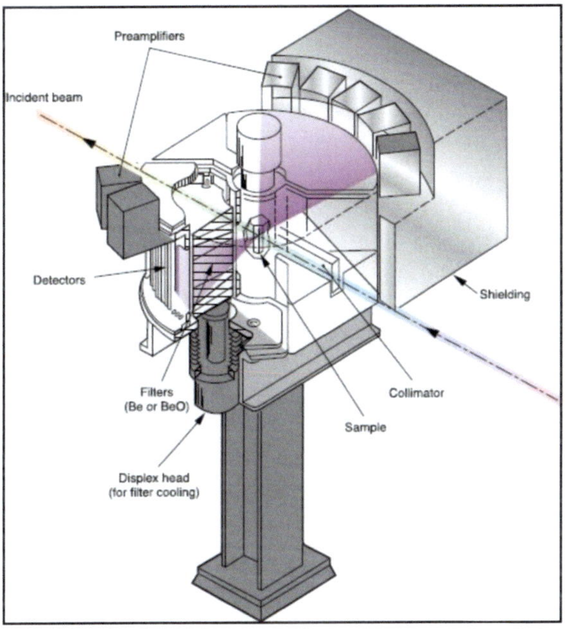

Figure 6. Schematic view of FDS at Lujan Center, LANL (Wavenumber from 0 to 2500 cm^{-1}).

Results and Discussions

Mineral Composition

Mineral compositions of the rock salt and its insoluble residue were determined using X-Ray Diffraction. The rock salt sample consists primarily of halite (NaCl) with minor amount of magnesite ($MgCO_3$). The insoluble residue consisting primarily of quartz and clays, including mica, smectite, and kaolinite, with trace amount of halite and magnesite.

Thermo Gravimetric Analysis (TGA)

The TGA results of the two samples are presented in Figure 7. The TGA analysis results indicate two weight loss steps for the insoluble residue at 54.1°C, and 449.4°C, with 4.09% and 19.63% weight percent respectively. The observed weight loss from room temperature to 54.1°C is likely due to release of physisorbed water presumably on clays. The second weight loss step at 449.4°C involves clay dehydration. Rock salt, which is intrinsically a multiphase assemblage, shows three weight loss steps: the first occurs at 51.1°C, the second at 94.0°C, and the last at 448.5°C. Similar to the insoluble residue, the observed weight loss (2.82%) from room temperature to 51.1°C is likely due to the release of physisorbed water on the clays.

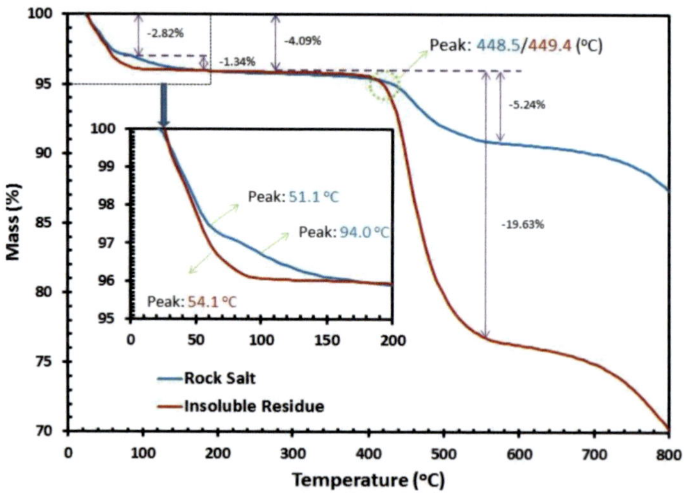

Figure 7. TGA analysis of rock salt and its insoluble residue. The moisture release of the halite phase is 0.07 weight%.

The next weight loss (1.34%) interval in rock salt from 51.1 to 150°C is in all likely hood due to release of water from decrepitated fluid inclusions as well as intergranular pore water confined in grain boundaries.

Finally at 449.4°C, similarly to the insoluble residue the weight loss (5.24%) occurs involving clay dehydration. Though TGA results assess the crucial temperature range of water release from salt, they fail in characterizing and quantifying the forms of water involved in the weight loss.

These TGA results resemble those published for halite (Zimmermann and Moretto, 1996, modified by the small amount of corrensite (Caporuscio et al., 2014. Pristine halite in our sample contains 0.07 (calculated from Figure 7. 0.07 = (2.84 + 1.34) - 4.09) weight% moisture (combined intergranular pore water and brine). The insoluble residue (0.5 weight% of the total sample) contains 22.5 weight% moisture (H_2O and OH), 4.09 weight% of which is released as H_2O by 150°C (Figure 7). The water content due to corrensite in our rock salt sample is 0.02 weight%.

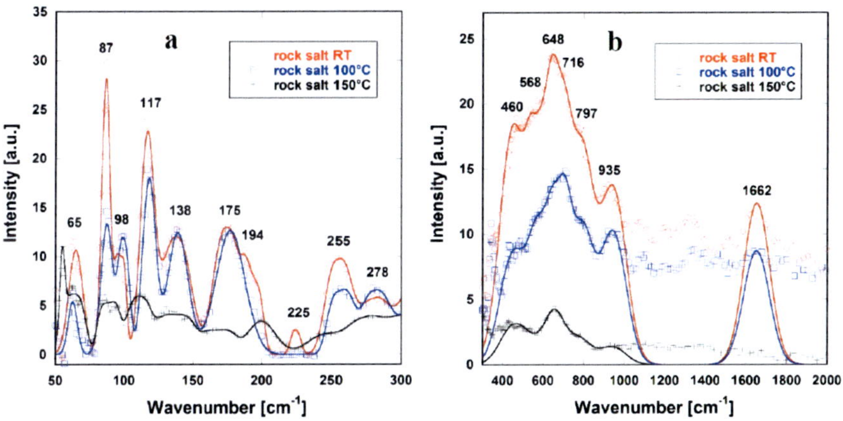

Figure 8. Three inelastic incoherent neutron scattering spectra of rock salt in (a) multiphonon domain, (b) librational and a small segment of the vibration domain at room temperature (RT, red), dehydrated at 100°C (blue), and at 150°C (black). The dots represent the experimental data; the lines are fitted results using Gaussians. The peak at 1662 cm^{-1} is the characteristic H_2O bending mode.

IINS Spectra of Rock Salt at Elevated Temperature

Figure 8 shows the IINS rock salt spectra of the room temperature (RT) sample and those after heat treatment at 100 and 150°C. The frequency in cm^{-1} (1 meV = 8.065 cm^{-1}) is given above each peak. Three frequency domains occur: (1) the multi-phonon modes-H2O between 1 and 300 cm^{-1} (Figure 8a),, 2) the water libration modes between 300-1100 cm^{-1} (Figure 8b), due to interactions between water molecules, and 3) the intramolecular vibrations from 1600 cm^{-1} upwards, the H-O-H bending mode of water around 1660 cm^{-1}.

Our focus is on the first two domains and their scattering intensity variation as a function of temperature. Though detailed assignment of the vibrational modes falls outside the scope of this study, some generalities may be observed. The first domain is characteristic for the interaction of moisture with NaCl lattice vibrational modes. There are six of these modes in NaCl, the highest energy, optical modes occurring at ~ 240 cm^{-1} (Raunio et al., 196). This domain has exceedingly narrow resonances with a Full Width at Half Maximum (FWHM) approximately half that of the narrowest and most intense translational peak in crystalline ice [ice_h] (Line and Kearly, 2000). It represents specific coupling between moisture confined at the intergranular interface and the halite phonons. They must be associated with grain boundaries for two reasons. First, the halite crystal structure has no sites to accommodate water molecules. Second the extreme narrowness of the resonances precludes coupling between different water molecules. Such modes are well-known for hydrogen atoms riding on catalyst supports, in which they follow adiabatically the plane modes in so-called riding modes (Mitchell et al., 2003, Mitchell et al, 2003, Cavallari et al., 2014). They have never been observed for moisture in zeolites or clays. We designate these modes “Chladni modes” to honor the father of acoustics Ernst Chladni (1756-1827). At higher frequencies, the multiphonon domain is followed by three resonances with somewhat broader FWHMs representing hydrogen stretches from intergranular moisture as well as fluid inclusions (Senesi et al., 2013).

The sudden rise in intensity around 300 cm^{-1} is designated the librational edge (LE). It marks the onset of the second domain representing interactions of water molecules with one another. These interactions are historically assigned to three modes around the center of mass of the water molecule: rock, twist, and wag, their energy sequencing being a matter of continuous debate as recently pointed out by Parker et al., (Parker et al.,

2011). This domain occurs as bands with or without structure in zeolites and clays (Ockwig et al., 2008, Line and Kearly, 2000, Neumann, 1995). It is characterized in our case by three high intensity resonances with a number of subsidiary ones with very large FWHM. Their intensity is a consequence of the large amplitude motion of librations, representing the motion of fluid in inclusions; their width reflects the multiple interactions of water with the multiple brine constituents and the 0.02 wt% moisture in corrensite.

What marks the utility of INNS is the difference in the response of the two spectral domains to heat treatment of the sample. After heat treatment of the sample at 100°C, the resonances in the first domain tend to coincide in position and intensity with those of the untreated sample, except for the two resonances at 65 and 225 cm^{-1}, which are possibly associated with corrensite dehydration The librational mode intensity in the second domain, on the other hand, is nearly halved together with a variation in the intensity ratio of the modes. It indicates primarily loss of the 0.02 wt% of moisture associated with corrensite that occurs between 65 and 75°C. Thus, the spectrum of the sample heat treated at 100°C shows the proportion of water confined in grain boundaries and that of brine included in fluid inclusions.

TGA results as well as the absence of the H_2O 1662 cm^{-1} bending mode in the at 150°C heat treated sample indicate beyond 150°C the absence of water in the sample, corrensite hydroxyl release taking off at 450°C. Accordingly the INNS spectra show virtually no resonances for the rock salt sample after treatment at this temperature.

IINS Spectra of the Insoluble Residue at Elevated Temperature

IINS measurement results of the insoluble residue at RT and after heat treatment at 150°C, and 450°C are presented in Figure 9. Our results show an almost identical spectrum for the RT and 150°C sample. On the other hand the sample heat treated at 450°C differs significantly. This is consistent with our TGA results, which indicate the onset of dehydroxylation of clays at~ 450°C.

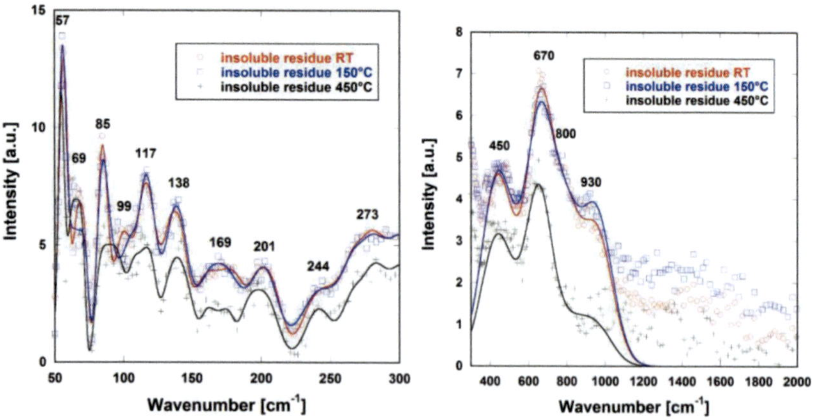

(a) (b)

Figure 9. Three inelastic incoherent neutron scattering spectra of the insoluble residue in (a) multiphonon domain, (b) librational domain at room temperature (RT, red), dehydrated at 150°C (blue), and at 450°C (black). The dots represent the experimental data; the lines are fitted results using Gaussians.

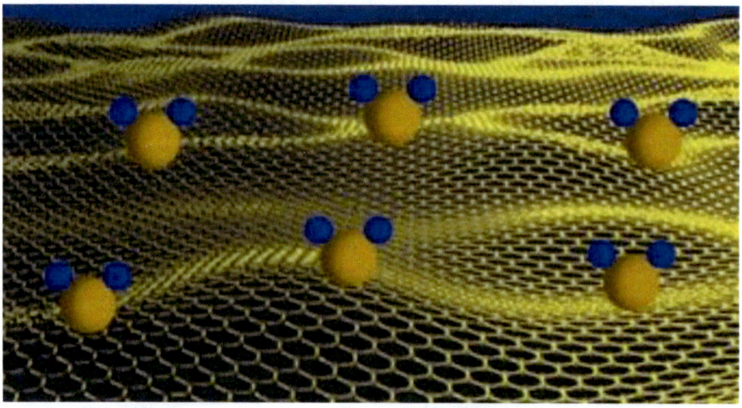

Figure 10. Image of water molecules riding on surface Chladni modes.

Moisture Release in the Multiphonon and Librational Domain

The temperature dependent release of water in the multiphonon scattering domain depends on the tenacity with which it is attached to the Chladni modes of the lattice (Figure 10). In the librational domain it depends on the strength of the walls of the fluid inclusions. To analyze these effects more precisely, we assume that at 150°C, all moisture has been released and the moisture-free 150°C halite spectrum is the same as the RT one. Thus, the difference spectrum between RT and 150°C is the total moisture contribution to the sample, i.e., the 0.07 weight% in pristine halite and 0.02 weight% in corrensite. The difference spectrum between 100°C and 150°C on the other hand is the contribution of the remaining moisture in the sample treated at 100°C. Figure 8 illustrates these two derived spectra fitted with an adequate number of Gaussians. Results are given in Table 1 indicating which modes release the most intergranular water or brine.

Table 1. Calculated peak area at corresponding frequencies and moisture release*

Frequency (cm^{-1})	Peak Area (RT-150°C)	% Water Remaining	Peak Area (100-150°C)	% Water Released	Form of Water
65	36	0	0	100	Intergranular pore water
87	158	39.87	63	60.13	
98	48	133.33	64	-33.33	
117	180	63.89	115	36.11	
139	180	73.89	133	26.11	
175	236	91.10	215	8.90	
194	63	22.22	14	77.78	
225	13	0.00	0	100.00	
255	135	49.63	67	50.37	
278	37	132.43	49	-32.43	
460	2454	39.89	979	60.11	Fluid inclusion
568	1163	51.94	604	48.07	
648	1574	50.95	802	49.05	
716	1058	61.25	648	38.75	
797	1563	56.81	888	43.19	
935	1778	71.88	1278	28.12	

*Caclulated weight percent loss per formula unit. Corrensite: 13.3. wt% H_2O at 65-75°C, 13.9 wt% OH at 450°C. Gypsum: 20.9 wt% H_2O at <100°C. Polyhalite: 6.0 wt% OH at ~250°C.Kaolinite: 26.4 wt% OH at >550°C

The multiphonon frequency range for NaCl varies between about 80 and 240 cm^{-1} with the soft modes being designated the acoustic modes and the most energetic ones the optic modes. Inspection of Table 1 indicates the mode dependent moisture release. The mode at 172 cm^{-1} represents the most tightly bound moisture. In general, the librational modes tend to release, as anticipated, more moisture over the considered temperature range relative to the multiphonon modes.

Conclusion

Our results show that it is possible with IINS to discriminate two moisture environments in halite due to a difference in response of the sample to heat treatment at 100°C. The first one, with very narrow resonances, is indicative of individual water molecules vibrating cooperatively with lattice vibrational modes of halite on grain boundaries. They are not affected by moderate sample heating. The second one is the result of fluid motion in brines which become visible once the signal due to corrensite moisture release is eliminated. The ability to partition the 0.07 weight % moisture in pristine halite over these two environments makes it possible to target the proper moisture moiety responsible for hydrolytic weakening of halite. In our open system, this moisture will be released to the surroundings. In a closed system the communication between brine released moisture, the source, and intergranular thin film growth, the sink, will show up in spectral intensity variations between the two domains. This spectral capability is a definitive advance urgently needed to place thermomechanical studies of halite on a firm footing and make pertinent robustness assessments of HNGW storage in bedded salt formations.

Acknowledgments

This work was funded by the Department of Energy, Office of Nuclear Energy, Used Fuel Disposition Campaign (UFD) program. IINS measurements were carried out at the Lujan Neutron Scattering Center at Los Alamos Neutron Science Center, which is funded by the Department of Energy's Office of Basic Energy Science. Emily Kluk of LANL carried out salt sample separation. Dr. B. Robinson, Director, Civilian Nuclear Program of Los Alamos National Laboratory, supported this initiative. Dr. J. A. TenCate from Los Alamos National Laboratory shared his expertise in physicist studies flow of earth material, particularly rock salt.

REFERENCES

Als-Nielsen, J. & McMorrow, D. (2000). *Elements of Modern X-ray Physics*. Wiley: New York.

Anovitz, L. M., Lynn, G. W., Cole, D. R., Rother, G., Allard, L. F., Hamilton, W. A., Porcar, L. & Kim, M. H. (2009). "A New Approach to Quantification of Metamorphism Using Ultra-Small and Small Angle Neutron Scattering." *Geochimica et Cosmochimica Acta.*, *73*, 7303-7324.

Aquilano, D., Otalora, F., Pastero, L. & Garcia-Ruiz, J. M. (2016). "Three Study Cases of Growth Morphology in Minerals: Halite, Calcite and Gypsum." *Prog. In Crystal Growth and Characterization of Materials*, *62*(2), 227-251.

Baas Becking, L. G. M. (1931). "Historical Notes on Salt and Salt Manufacture." *The Scientific Monthly*, *32*, 434-46.

Baas Becking, L. G. M. (2015). *Baas Becking's: Geobiology or Introduction to Environmental Science*. D.E. Canfield (ed.), Chichester (UK) and Hoboken (USA): Wiley Blackwell.

Baumgartner, M. & Bakker, R. J. (2010). "Raman spectra of ice and salt hydrates in synthetic fluid inclusions." *Chemical Geology*, *275*, 58-66.

Bloch, M. R. (1963). "The Social Influence of Salt." *Scientific American*, *209*, 89-98.

Börner, K. (2003). *Evaporite basins with emphasis on the Permian Zechstein.* PDF Abstract.

Bradshaw, R. L. & Sanchez, F. (1969). Migration of brine cavities in rock salt, *Journal of Geophysical Research*, *74*(17), 4209-4212.

Braitsch, O. (1971). *Salt Deposits: Their Origin and Composition*. New York: Springer-Verlag.

Caporuscio, F. A., Boukhalfa, H., Cheshire, M. C. & Ding, M. (2014). "Brine Migration Experimental Studies for Salt Repositories." *Fuel Cycle Research & Development*, Los Alamos National Laboratory Report, LA-UR-14-26603.

Carter, N. L. & Hansen, F. D. (1983). "Creep of Rocksalt." *Tectonophysics*, *92*, 275-333.

Cheung, H., Fuller, M. E. & Gaffney, E. G. (1980). "Modeling of Brine Migration in Halite." In *Scientific Basis for Nuclear Waste Management*, edited by Northrup, J.M., 2: 471-478, New York: Plenum Press.

Cole, D. R., Herwig, K. W., Mamontov, E. & Larese, J. Z. (2006). "Neutron Scattering and Diffraction Studies of Fluids and Fluid-Solid Interactions, *Rev. in Min. & Geochem.*, *63*, 313-362.

Cowley, J. M. (1975). *Diffraction Physics.* North-Holland: Amsterdam.

De Jong, H. N. & Conte, T. (2017). "Seasonal and Prehistoric Salt Consumption Through $^{87}Sr/^{86}$ Sr Analysis of Tooth Enamel. *Goldschmidt Abstracts*, *875*.

Denton, D. (1982). *The Hunger for Salt. An Anthropological, Physiological and Medical Analysis*. Berlin Heidelberg New York: Springer-Verlag.

Desbois, G., Urai, J. L. & De Bresser, H. P. (2012). "Fluid Distribution in Grain Boundaries of Natural Fine-Grained Rock Salt Deformed at Low Differential Stress (Qom Kuh Salt Formation, Central Iran): Implications for Rheology and Transport Properties." *Journal of Structural Geology*, 43, 128-143.

Ding, M., Hartl, M., Xu, H., Hjelm, R. P., Miller, A. & Wang, Y. (2013). "Pore Characteristics and Their Evolution in Clays Using Small-Angle Neutron Scattering." *In Proceedings, 2013 International High-Level Radioactive Waste Management*. Albuquerque, New Mexico.

Ding, M., Watkins, E., Hartl, M. & Daemen, L. (2015). "Water Signatures and Their Thermal Stability in Bedded Salt for Nuclear Waste Storage: An Incoherent Inelastic Neutron Spectroscopy Study." *Environ. Sci. Technol. Lett.*, *2* (11), 308–313.

Dore, J. C. (2000). "Structural Studies of Water in Confined Geometry by Neutron Diffraction." *Chem. Phys.*, *258*, 327-347.

Fayer, M. D. & Levinger, N. F. (2010). "Analysis of water in confined geometries and at interfaces." *Annual Review Analytical Chemistry*, *3*, 89-107.

Flad, R. K. (2011). *Salt Production and Social Hierarchy in Ancient China: An Archaeological Investigation of Specialization in China's Three Gorges*. New York.: Cambridge University Press.

Gu, X., Mildner, D. F. R., Cole, D. R., Rother, G., Slingerland, R. & Brantley, S. L. (2016). "Quantification of Organic Porosity and water Accessibility in Marcellus Shale Using Neutron Scattering." *Energy & Fuels*, *30*(6) 4438-4449.

Haak, W., Brandt, G., de Jong, H. N., Meyer, C., Ganslmeier, R., Heyd, V., Hawkesworth, C., Pike, A. W. G., Miller, H. & Alt, K. W. (2008). "Ancient DNA, Strontium Isotopes and Osteological Analyses Shed Light on Social and Kinship Organization of the Later Stone Age." *PINAS*, *105*, 18226-18231.

Hansen, F. D. & Leigh, C. D. (2011). *"Salt Disposal of Heat-Generating Nuclear Waste."* SAND2011-0161, Albuquerque: Sandia National Laboratories Report.

Hoff, J. H. V., van 't. (1905). "Zur Bildung der Ozeanischen Salzablagerungen. Z." *Anorg. Chem.*, *47*(1), 244-280.

Houben, M. E., ten Hove, A., Peach, C. J. & Spiers, C. J. (2013). "Crack Healing in Rock Salt via Diffusion in Adsorbed Aqueous Films: Microphysical Modeling versus Experiments." *Phys. Chem. Earth*, *64*, 95-104.

Hwang, Y., Chambré, Pigford, T. H. & Lee, W. W. L. (1990). "Brine Migration in a Salt Repository, Radioactive Waste Management." *Nuclear Technology*, *90*, 205-214.

Jordan, A. B., Boukhalfa, H., Caporuscio, F. A., Robinson, B. A. & Stauffer, P. H. (2015). "Hydrous Mineral Dehydration Around Heat-Generating Nuclear Waste in Bedded Salt Formations." *Environ. Sci. & Technol.*, *49*, 6783-6790.

Knauth, L. P. & Kumer, M. B. (1981). "Trace Water Content of Salt in Louisiana Salt Domes." *Science*, *213*, 1005-1007.

Leitner, C., Neubauer, F., Marschallinger, R., Genser, J. & Bernroider, M. (2013). "Origin of Deformed Halite Hopper Crystals, Pseudomorphic Anhydrite Cubes and Polyhalite in Alpine Evaporites (Austra, Germany)" *Int. J. Earth Sc.*, *102*(3), 813-829.

Li, J. C., Londono, D. K., Ross, D. K., Finny, J. L., Tomkinson, J. & Sherman, W. F. (1991). An inelastic incoherent neutron scattering study of ice II, IX, V, and VI in the range from 2 to 140 meV. *J. Chem. Phys.*, *94*, 6770-6775.

Llewellyn, P. G. (1968). "Dendritic Halite Pseudomorphs from the Keuper Marl of Leicestershire England." *Sedimentology*, *11*, 293-297.

Line, C. M. B. & Kearly, G. J. (2000). An inelastic incoherent neutron scattering study of water in small-pored zeolites and other water bearing minerals. *J. Chem. Phys.*, *112* (20), 9058-9067.

Lowenstein, T. K., Timofeeff, M. N., Brennan, S. T., Hardie, L. A. & Demicco, R. V. (2001). "Oscillations in Phanerozoic Seawater Chemistry: Evidence from Fluid Inclusions." *Science*, *294*, 1086-1088.

Maccarini, M. (2007). "Water at solid surfaces: A review of selected theoretical aspects and experiments on the subject." *Biointer Phases*, *2*(3), MR1-MR15.

Mitchell, P. C. H., Ramirez-Cuesta, A. J., Parker, S. F. & Tomkinson, J. (2003). "Inelastic Neutron Scattering in Spectroscopic Studies of Hydrogen on Carbon-Supported Catalysts-Experimental Spectra and Computed Spectra of Model Systems." *J. Molec. Struct.*, *107*, 6838-6845.

Muhammed, N. (2015*). "Deformation and Transport Processes in Salt Rocks: An Experimental Study Exploring Effects of Pressure and Stress Relaxation."* PhD diss., Utrecht University.

Multhauf, R. P. (1978). *Neptune's Gift: a History of Common Salt*. Johns Hopkins: Univ. Press.

Nenquin, J. (1961). *Salt: A Study in Economic Prehistory*. De Tempel, Brugge.

Neumann, D. A. (1995). Quasielastic and inelastic neutron scattering of layered silicates. In *Access in nanoporous materials*. Pinnavaia, T.J., Thorpe, M.F., Eds. Plenum Press., 381-404

Ochsenius, C. (1877). *The formation of rock salt deposits and their mother liquors salts under special consideration of the roaches of Douglas hall in the Egeln'schen hollow*, Halle C.E.M. Pfeffer.

Ochsenius, C. (1906). *Theories about the origin of salt deposits in Germany's potash industry supply to newspaper industry* 2nd ed, 1-8

Ockwig, N. W., Cygan, R. T., Hartl, M. A., Daemen, L. L. & Nenoff, T. M. (2008). "Incoherent Inelastic Neutron Scattering Studies of Nanoconfined Water in Clinoptilolite and Heulandite Zeolites." *Journal of Physical Chemistry.*, *112*, 13629-13634.

Olander, D. R., Machiels, A. J. & Yagnik, S. K. (1981). "Thermomigration of two-phase inclusions in salt" *Trans. Am. Nucl. Soc.*, *35*, 168-169.

Parker, S. F., Refson, K., Bewley, R. L. & Dent, G. (2011). "Assignment of the vibrational spectra of lithium hydroxide monohydrate." *J. Chem. Phys.*, *134*, 84503-84512.

Peach, C. J. & Spiers, C. J. (1995). "Influence of Crystal Plastic Deformation on Dilatancy and Permeability Development in Synthetic Salt Rock." *Tectonophysics*, *256* (1-4), 101-128.

Pechar, F., Schweiss, P. & Fuess, H. (1982). Inelastic neutron scattering study of low frequency vibrations of water molecules in Natrolite. *Chemické Zvesti*, *36*(6) 779-783.

Pigford, T. H. (1982). "Migration of Brine Inclusions in Salt." *Nuclear Technology*, *56*, 93-101.

Rahimpour-Bonab, H., Shariatinia, Z. & Siemann, M. G. (2007). "Origin and Geochemistry of Miocene Marine Evaporates Associated with Red Beds: Great Kazir Basin." Central Iran. *Geol. J.*, *42*, 37-54.

Radlinski, A. P., Boreham, C. J., Wignall, G. D. & Lin, J. S. (1996). "Microstructure Evolution of Source Rocks during Hydrocarbon

Generation: A Small-Angle Scattering Study." *Physical Review B.*, *53*, 14152-14160.

Raunio, G., Almqvist, L. & Stedman, R. (1969). "Phonon Dispersion Relations in NaCl." *Phys. Rev.*, *178*(3) 1496-1501.

Robinson, B. A., Elkins, N. Z. & Carter, J. T. (2012). "Development of a U.S. Nuclear Waste Repository Research Program in Salt." *Nuclear Technology*, *180*, 122-138.

Roedder, E. & Belkin, H. E. (1979). "Application of Studies of Fluid Inclusions in Permian Salado Salt, New Mexico, to the Problems of Siting the Waste Isolation Pilot Plant" In: *Scientific Basis for Nuclear Waste Management*, edited by McCarthy, G.J., 453-464. New York: Plenum Press.

Roedder, E. & Belkin, H. E. (1980). "Thermal Gradient Migration of Fluid Inclusions in Single Crystals of Salt from the Waste Isolation Pilot Plant Site (WIPP)." In *Scientific Basis for Nuclear Waste Management.* Edited by Northrup, J.M., *2*, 453-464, New York: Plenum Press.

Roedder, E. & Basset, R. L. (1981). "Problems in Determination of the Water Content of Rock-Salt Samples and Its Significance in Nuclear-Waste Storage Siting." *Geology*, *6*, 525-530.

Roedder, E. (1984). "The fluids in salt." *American Mineralogist*, *69*, 413-439.

Senesi, R., Flammini, D., Kolesnikov, A. L., Murray, E. D., Galli, G. & Andreani, C. (2013). The quantum mature of the OH stretching mode in ice and water probed by neutron scattering experiments. *J. Chem. Phys*, *139*, 074504.

Schoenherr, J., Urai, J. L., Kukla, P. A., Littke, R., Schleder, Z., Larroque, J. M., Newall, M. J., Al-Abry, N., Al-Siyahi, H. A. S. & Rawahi, Z. (2007). "Limits to the Sealing Capacity of Rock Salt: A Case Study of the Infra-Cambrian Ara Salt from the South Oman Salt Basin." *AAPG Bull*, *91*(11), 1541-1557.

Sears, V. F. (1992). "Neutron Scattering Lengths and Cross Sections." *Neutron News*, *3*(3), 26-37.

Spiers, C. J., Urai, J. L. & Lister, G. S. (1988). "The Effect of Brine (Inherent or Added) on Rheology and Deformation Mechanisms in Salt Rock." In: *The Mechanical Behaviour of Salt II*, edited by Hardy, H. R. Jr., and Langer eds., M. Trans. Tech., 89–102, Clausthal-Zellerfeld, German.

Sun, X. H., Hu, M. Y., Liu C. L., Jiao, P. C., Ma, L. C., Wang, X. & Zhan, X. C. (2013). "Composition Determination of Single Fluid Inclusions in Salt Minerals by Laser Ablation ICP-MS." *Chinese Journal of Analytical Chemistry*, *41*(2), 235-241.

Tamman, G. (1903). *Crystallizing and melting*. Nabu Press.

Tarjus, G., Diraison, M., Leicknam, J. Cl. & Bratos, S. (1993). "Inelastic Neutron Scattering from Liquid Water. Theory and Simulation." *Journal of Molecular Structure.*, *296*, 199-204.

Timofeeff, M. N., Lowenstein, T. K., Brennan, S. T., Demicco, R. V., Zimmermann, H., Horita, J. & von Borstel, L. E. (2001). "Evaluating seawater chemistry from fluid inclusions in halite: Examples from modern marine and non-marine environments." *Geochim. Cosmochim. Acta*, *65*(14), 2293-2300

Urai, J. L. (1983). "Water Assisted Dynamic Recrystallization and Weakening in Polycrystalline Bischofite." *Tectonophysics*, *96*, 125-157.

Urai J. L., Spiers, C. J., Zwart, H. J. & Lister, G. S. (1986). "Weakening of Rock Salt by Water During Long-Term Creep." *Nature*, *324*, 554-557.

Urai, J. L. & Spiers, C. J. (2007). "The Effect of Grain Boundary Water on Deformation Mechanisms and Rheology of Rock Salt during Long-Term Deformation." *In Proceedings of the 6th conference on the mechanical behavior of salt*, *'SaltMech6'*, edited by M. Wallner, K. Lux, W. Minkley and H. Hardy Jr., 22-25. Hannover, Germany.

Usiglio, J. (1849). Analysis of the water of the Mediterranean on the coasts of France. *Annal. Chim. et Phys.*, XXVII 92-107, 172-191.

Wald, M. L. (2014). "Nuclear Waste Solution Seen in Desert Salt Beds." *The New York Times*, Feb., *9*, 2014.

Willis, B. T. M. & Carlile, C. J. (2009). *Experimental Neutron Scattering.* Oxford University Press: New York.

Zimmerman, J. L. & Moretto, R. (1996). Release of water and gasses from halite crystals. *Eur. J. Mineral.*, *8*, 413-422.

In: Solution Chemistry
Editor: Yongliang Xiong
ISBN: 978-1-53613-101-7

Chapter 4

ABSENCE OF MINERAL COLLOIDS IN HIGH IONIC STRENGTH SOLUTIONS ASSOCIATED WITH SALT FORMATIONS: EXPERIMENTAL DETERMINATION AND APPLICATIONS FOR NUCLEAR WASTE MANAGEMENT

Yongliang Xiong[1,*], Leslie Kirkes[1], Sungtae Kim[1], Cassie Marrs[1], Jandi Knox[1], Justin Dean[1], Haoran Deng[2] and Martin Nemer[2,*]

[1]Sandia National Laboratories (SNL), Carlsbad Programs Group, Carlsbad, NM, US

[2]Sandia National Laboratories (SNL), Albuquerque, NM, US

ABSTRACT

Radionuclides and heavy metals easily sorb onto colloids. This phenomenon can have a beneficial impact on environmental clean-up activities if one is trying to scavenge hazardous elements from soil for example. On the other hand, it can have a negative impact in cases where

* Corresponding Author Email: yxiong@sandia.gov.

one is trying to immobilize these hazardous elements and keep them isolated from the public. Such is the case in the field of radioactive waste disposal. Colloids in the aqueous phase in a radioactive waste repository could facilitate transport of contaminants including radioactive nuclides. Salt formations have been recommended for nuclear waste isolation since the 1950's by the U.S. National Academy of Science. In this capacity, salt formations are ideal for isolation of radioactive waste. However, salt formations contain brine (the aqueous phase), and colloids could possibly be present. If present in the brines associated with salt formations, colloids are highly relevant to the isolation safety concept for radioactive waste. The Waste Isolation Pilot Plant (WIPP) in southeast New Mexico is a premier example where a salt formation is being used as the primary isolation barrier for radioactive waste. WIPP is a U.S. Department of Energy geological repository for the permanent disposal of defense-related transuranic (TRU) waste. In addition to the geological barrier that the bedded salt formation provides, an engineered barrier of MgO added to the disposal rooms is used in WIPP. Industrial-grade MgO, consisting mainly of the mineral periclase, is in fact the only engineered barrier certified by the U.S. Environmental Protection Agency (EPA) for emplacement in the WIPP. Of interest, an $Mg(OH)_2$-based engineered barrier consisting mainly of the mineral brucite is to be employed in the Asse repository in Germany. The Asse repository is located in a domal salt formation and is another example of using salt formations for disposal of radioactive waste. Should colloids be present in salt formations, they would facilitate transport of contaminants including actinides. In the case of colloids derived from emplaced MgO, it is the hydration and carbonation products that are of interest. These colloids could possibly form under conditions relevant in particular to the WIPP. In this chapter, we report a systematic experimental study performed at Sandia National Laboratories in Carlsbad, New Mexico, related to the WIPP engineered barrier, MgO. The aim of this work is to confirm the presence or absence of mineral fragment colloids related to MgO in high ionic strength solutions (brines). The results from such a study provides information about the stability of colloids in high ionic strength solutions in general, not just for the WIPP. We evaluated the possible formation of mineral fragment colloids using two approaches. The first approach is an analysis of long-term MgO hydration and carbonation experiments performed at Sandia National Laboratories (SNL) as a function of equivalent pore sizes. The MgO hydration products include $Mg(OH)_2$ (brucite) and $Mg_3Cl(OH)_5 \cdot 4H_2O$ (phase 5), and the carbonation product includes $Mg_5(CO_3)_4(OH)_2 \cdot 4H_2O$ (hydromagnesite). All these phases contain magnesium. Therefore, if mineral fragment colloids of these hydration and carbonation products were formed in the SNL experiments mentioned above, magnesium concentrations in the filtrate from the experiments would show a dependence on ultrafiltration. In other words, there would be a decrease in magnesium concentrations as a function of ultrafiltration with decreasing molecular weight (MW) cut-offs for the filtration. Therefore, we performed ultrafiltration on solution samples from the SNL hydration and carbonation experiments as a function of equivalent pore size. We filtered solutions using a series of MW cut-off filters at 100 kD, 50 kD, 30 kD and 10 kD. Our results demonstrate that the magnesium concentrations remain constant with decreasing MW cut-offs, implying the absence of mineral fragment colloids. The second approach uses spiked Cs^+ to indicate the possible presence of mineral fragment colloids. Cs^+ is easily absorbed by colloids. Therefore, we added Cs^+ to a subset of SNL MgO hydration and carbonation experiments. Again, we filtered the solutions with a series of MW cut-off filters at 100 kD, 50 kD, 30 kD and 10 kD. This time we measured the concentrations of Cs. The concentrations of Cs do not change as a function of MW cut-offs, indicating the absence of colloids from MgO hydration and carbonation products. Therefore, both approaches demonstrate the absence of mineral fragment colloids from MgO hydration and carbonation products. Based on our experimental results, we acknowledge that mineral fragment colloids were not formed in the SNL MgO hydration and carbonation experiments, and we further conclude that high ionic strength solutions associated with salt formations prevent the formation of mineral fragment colloids. This is due to the fact that the high ionic strength solutions associated with salt formations have high concentrations of both monovalent and divalent metal ions that are orders of magnitude higher than the critical coagulation concentrations for mineral fragment colloids. The absence of mineral fragment colloids in high ionic strength solutions implies that contributions from mineral fragment colloids to the total mobile source term of radionuclides in a salt repository are minimal.

Keywords: Nuclear Waste Management, colloidal transport, environmental chemistry, engineered barrier, Waste Isolation Pilot Plant (WIPP)

INTRODUCTION

A colloid is a dispersed, separable phase that is suspended within a second phase, in this case, aqueous brine solutions. In this study, we follow the operational terminology defined by Ross and Sherrell (1999) regarding colloidal and dissolved fractions; i.e., the colloidal fraction can be separated from solution using filters/ultrafilters with pore sizes ranging from 10 kDa to 0.2 μm.

Colloids can facilitate transport of contaminants including actinides (e.g., Silva and Nitsche, 1995; Wen et al., 1999), and therefore, the possible persistence of colloids is important in performance of the isolation characteristic of a nuclear waste repository. The persistence of colloids may increase actinide mobility and enhance transport of actinides.

Salt formations are ideal for nuclear waste isolation, as recommended by the National Academy of Science as early as the 1950's (U.S. National Academy of Science, 1957), as salt formations have the following advantages: (1) low permeability; (2) plastic deformation after closure of a repository to encapsulate wastes, (3) self-healing of fractures, and (4) high thermal conductivity of salt.

There are four types of colloids considered in the conceptual model for the Waste Isolation Pilot Plant (WIPP), a nuclear waste repository in a bedded salt formation, located in southeast New Mexico. These are: intrinsic colloids, mineral fragment colloids, humic colloids, and biocolloids. Intrinsic colloids are generated by polymerization of actinides via hydrolysis. When actinide elements (intrinsic colloids or aqueous species) are absorbed onto inorganic or organic aquatic colloids because of their affinity to these colloids, the final products become actinide pseudocolloids (e.g., Olofsson and Allard, 1986). Mineral fragment colloids are pseudocolloids formed by sorption of aqueous actinide species or intrinsic actinide colloids onto inorganic aquatic colloids (Zhao and Steward, 1997). Aquatic colloids that can absorb aqueous actinide species or intrinsic actinide colloids may be inorganic or organic. Humic substance is an example of aquatic organic colloids (Kim et al., 1994). Various mineral fragments, such as α-Al_2O_3 (Olofsson et al., 1983), bentonite (Kelly et al., 1996), hematite and goethite (Itagaki et al., 1991; Kelly et al., 1996; Wang et al., 2015), montmorillonite (Olofsson and Allard, 1986); SiO_2 (Allen and Matijević, 1969, 1971; Olofsson and Allard, 1986; Itagaki et al., 1991), and ZrO_2 (Bitea et al., 2003), are examples of aquatic inorganic colloids.

The stability of both intrinsic colloids and pseudocolloids, highly depends on the ionic strength of the aqueous solution (Allen and Matijević, 1969, 1971; van Olphen, 1991). The ionic strength at which rapid destabilization and agglomeration of both intrinsic colloids and pseudocolloids occurs is referred to as the critical coagulation concentration, c.c.c. Regarding the c.c.c. required for mineral fragment colloids, electrolytes containing monovalent cations such as Na^+ and K^+ have a higher c.c.c. up to 0.15 M (van Olphen, 1991). In contrast, electrolytes containing divalent cations such as Mg^{2+} and Ca^{2+} have a much lower c.c.c., ranging from 5×10^{-4} to 2×10^{-3} M (van Olphen, 1991).

Solutions associated with salt formations are characterized by high ionic strengths. For example, in the WIPP, there are two brines considered for the performance of the repository, i.e., Generic Weep Brine (GWB), and Energy Research and Development Administration (WIPP Well) 6 (ERDA-6). They have ionic strengths of 8.26 mol•kg^{-1} and 5.82 mol•kg^{-1}, respectively (Xiong and Lord, 2008). In the Asse, the brine used for assessing performance is the Q-brine, an Mg-Cl dominated brine with an ionic strength of ~13 mol•kg^{-1} (Schuessler et al., 2001). In such high ionic strength solutions with multiple components, it is expected that the formation of mineral fragment colloids is unlikely, but has not been experimentally demonstrated.

In work performed by Altmaier et al., (2004), the authors suggested the potential formation of mineral fragment colloids composed of magnesium chloride hydroxide hydrate in pure $MgCl_2$ solutions. There are two types of magnesium chloride hydroxide hydrates that are typically observed, i.e., $Mg_2Cl(OH)_3$•$4H_2O$ (called phase 3 in the cement industry) and $Mg_3Cl(OH)_5$•$4H_2O$ (called phase 5 in the cement industry). Altmaier et al., (2004) concluded that Th(IV) could adsorb to phase 3 mineral fragments to form actinide pseudocolloids.

In the WIPP near field geochemical model there are two types of brines, ERDA-6 and GWB. GWB is a Na-Mg-Cl dominated brine, and ERDA-6 is a Na-Cl dominated brine. These brines are characterized by high ionic strengths, with GWB and ERDA-6 having ionic strengths of

8.26 mol•kg^{-1} and 5.82 mol•kg^{-1}, respectively (Xiong and Lord, 2008). These brines are multi-component, and are significantly different from pure $MgCl_2$ solutions used in the experiments of Altmaier et al., (2004). These brines have high concentrations of both monovalent and divalent ions. For example, the ERDA-6 brine has 5.35 mol•kg^{-1}, 0.106 mol•kg^{-1}, 0.0209 mol•kg^{-1}, 0.0132 mol•kg^{-1} for Na^+, K^+, Mg^{2+}, and Ca^{2+} (Xiong and Lord, 2008), respectively. Similarly, GWB has 4.04 mol•kg^{-1}, 0.532 mol•kg^{-1}, 1.16 mol•kg^{-1}, 0.0163 mol•kg^{-1} for Na^+, K^+, Mg^{2+}, and Ca^{2+} (Xiong and Lord, 2008), respectively. These concentrations are much higher than the c.c.c. mentioned above, implying that any mineral fragments and their pseudocolloids are unstable in these brines. The presence of both monovalent and divalent cations in the WIPP brines is important. In fact, Altmaier et al., (2004) mentioned that "in NaCl or dilute $MgCl_2$ solutions, neither solid $Mg_2(OH)_3Cl\cdot 4H_2O(s)$ nor its colloids are stable." Therefore, the chemical compositions of the brine solutions make it unlikely that magnesium chloride hydroxide hydrate colloids would form (Xiong et al., 2010a), but to this date the stability of these colloids has not been demonstrated.

When industrial grade MgO is added as an engineered barrier to react with GWB, only phase 5 is formed (Xiong and Lord, 2008). However, when the ERDA-6 brine reacts with industrial grade MgO, neither phase 5 nor phase 3 is observed (Xiong and Lord, 2008). These observations are in agreement with thermodynamic calculations (Xiong et al., 2010b). Therefore, it has been reasoned that colloidal suspensions of either phase 3 or phase 5 are not a concern in the ERDA-6 brine. In the presence of $CO_2(g)$, when GWB and ERDA-6 react with MgO, hydromagnesite forms (Xiong and Lord, 2008).

The objective of this work is to experimentally evaluate the possible presence of mineral fragment colloids in high ionic strength solutions associated with salt formations, with a special reference to the possible presence of mineral fragment colloids resulting from the reaction of MgO with WIPP brines, including hydration and carbonation products. The results from this study will also have broad applications to other geological repositories in salt formations in general.

METHODS

Experimental Design

In this work, we investigated the possible presence of mineral fragment colloids by using the method of ultrafiltration. We designed two sets of experiments for this purpose. In the first set of experiments, we withdraw five aliquots of solution samples from long-term MgO hydration or carbonation experiments, which were set-up in 2008 in various solutions including synthetic GWB and ERDA-6 (see Tables 1 and 2 for details) at Sandia National Laboratories (SNL) in Carlsbad, NM (Deng, Nemer, Xiong, 2006; Deng, Xiong, Nemer, Johnsen, 2009). Then, each aliquot is separately subject to phase separation with filtration and ultrafiltration corresponding to 0.2 μm and molecular weight (MW) cut-off filters at 100 kD, 50 kD, 30 kD and 10 kD. Then, these samples were analyzed for Mg using inductively coupled plasma atomic emission spectrometer (ICP-AES).

Table 1. Recipes for Synthetic Salado and Castile Brines on Molarity Scale (from Xiong, 2008)

	GWB	ERDA-6
Salt	Grams for1L	Grams for 1L
$Na_2B_4O_7{\bullet}10H_2O$	15.06	6.00
NaBr	2.74	1.13
LiCl	0.19	None
NaCl	179.61	261.64
KCl	34.84	7.23
$MgCl_2{\bullet}6H_2O$	207.05	3.86
$CaCl_2{\bullet}2H_2O$	2.03	1.76
Na_2SO_4	25.23	23.70

Theoretically, if colloids are present, we would expect to see a significant drop in Mg concentrations after filtration and a consistently lower concentration of Mg in successive filtrations as Mg-bearing colloids are retained by filtration, as the hydration products, $Mg(OH)_2$ (brucite), and $Mg_3Cl(OH)_5 \cdot 4H_2O$ (phase 5), and the carbonation product, $Mg_5(CO_3)_4(OH)_2 \cdot 4H_2O$ (hydromagnesite), contain magnesium.

Table 2. Recipes for Synthetic Salado and Castile Brines on Molality Scale (from Xiong, 2008)

	GWB	ERDA-6
Salt	Grams for 1,000 g H_2O	Grams for 1,000 g H_2O
$Na_2B_4O_7 \cdot 10H_2O$	16.96	6.73
NaBr	3.09	1.27
LiCl	0.21	None
NaCl	202.32	293.38
KCl	39.25	8.11
$MgCl_2 \cdot 6H_2O$	233.23	4.33
$CaCl_2 \cdot 2H_2O$	2.29	1.97
Na_2SO_4	28.42	26.58

In the second set of experiments, we introduced the tracer, a CsCl solution, into solution samples taken from the same long-term experiments (Deng, Nemer, Xiong, 2006; Deng, Xiong, Nemer, Johnsen, 2009), to detect the presence of colloids. It has been repeatedly demonstrated that Cs^+ is easily absorbed onto various colloids (e.g., Bascetin et al., 2003; Chen et al., 2005; Cheng and Saiers, 2010, 2015; Apalkov et al., 2016). Again, we withdraw five aliquots of solution samples with CsCl added. Then, each aliquot is separately subject to phase separation with filtration and ultrafiltration corresponding to 0.2 μm and MW cut-off filters at 100 kD, 50 kD, 30 kD and 10 kD. After that, solution samples were analyzed for Cs using inductively coupled plasma mass spectrometer (ICP-MS). Theoretically, if mineral fragment colloids are present, we would expect to see a significant drop in Cs concentrations as a function of decreasing equivalent pore sizes, because mineral fragment colloids with Cs^+ absorbed onto them are retained by filtration and ultrafiltration.

If we observe a decrease in Mg and Cs concentrations as a function of decreasing equivalent pore sizes in these experiments, it indicates the presence of mineral fragment colloids. Conversely, if there is no dependence in Mg and Cs concentrations on decreasing equivalent pore sizes, i.e., Mg and Cs concentrations remain constant regardless of equivalent pore sizes, it indicates the absence of any mineral fragment colloids.

Analytical Equipment and Experimental Material

Ultrafiltration devices, Amicon® Ultra centrifugal filters, for this work are from MilliporeSigma Company. The ultrafiltration devices used in experiments include MW cut-off ultrafilters at 100 kD, 50 kD, 30 kD and 10 kD, corresponding to equivalent pore sizes of 7, 5, 4, and 3 nanometers (nm), respectively. The centrifuge used in experiments for ultrafiltration was an Avanti® J-E centrifuge from Beckman Coulter Company.

Elemental concentrations of magnesium were determined by using the Perkin Elmer Optima DV 3300 inductively coupled plasma atomic emission spectrometer (ICP-AES). The Perkin Elmer NexIon 300D inductively coupled plasma mass spectrometer (ICP-MS) is being used for determination of elemental concentrations of cesium.

The industrial grade MgO used in long-term hydration and carbonation experiments were from Martin Marietta Magnesia Specialties LLC (Deng et al., 2008, 2009). The distributions of particle sizes for the industrial grade MgO were as follows: > 2 mm, 7.02 ± 0.91 wt%; 1 mm—2 mm, 32.52 ± 1.78 wt%; 600 μm—1 mm, 20.25 ± 1.28 wt%; 300 μm—600 μm, 12.74 ± 2.19 wt%; 150 μm—300 μm, 5.35 ± 0.70 wt%; 75 μm—150 μm, 3.36 ± 0.35 wt%; and < 75 μm, 17.91 ± 1.88 wt% (Deng et al., 2008). The chemical compositions of the industrial grade MgO were 98.46 ± 2.54 wt% of MgO; 0.87 ± 0.03 wt% of CaO; 0.13 ± 0.018 wt% of Al_2O_3; 0.12 ± 0.01 wt% of total iron oxide; and 0.31 ± 0.01 wt% of SiO_2 (Deng et al., 2008). The MgO with three particle sizes were used in experiments: (1) as-received MgO (mixed size MgO); (2) large particle size MgO (1 mm—2 mm); and small particle size MgO (< 75 μm).

The GWB and ERDA-6 used in long-term hydration and carbonation experiments were synthesized according to the recipes detailed in Tables 1 and 2 (Xiong, 2008). The chemicals used for making the synthetic GWB and ERDA-6 were reagent grade chemicals from Fisher Scientific. In addition, 1.0 M $MgCl_2$ solution was used in long-term hydration experiments, and simplified GWB (1.0 M $MgCl_2$ + 3.6 M NaCl) was used for both long-term hydration and carbonation experiments.

Experimental Results

Magnesium Concentrations as a Function of Equivalent Pore Sizes

In this section, magnesium concentrations as a function of equivalent pore size for samples taken from the SNL MgO long-term hydration and carbonation experiments in GWB and ERDA-6 are presented. It should be mentioned that the hydration products for MgO hydration experiments in GWB are phase 5 and brucite, whereas the hydration product in ERDA-6 is brucite (Xiong and Lord, 2008). The carbonation product for MgO carbonation experiments in both GWB and EDRA-6 is hydromagnesite(5424) (Xiong and Lord, 2008). In MgO carbonation experiments, hydration products characteristic of GWB and ERDA-6, as mentioned before, are also present.

In addition, the hydration products for MgO long-term hydration experiments with a 1.0 M $MgCl_2$ solution are brucite and phase 5. The hydration and carbonation products for the experiments with the simplified GWB are identical to those with GWB.

In Figures 1 and 2, magnesium concentrations as a function of ultrafiltration in terms of equivalent pore size in nanometer for long-term MgO hydration experiments in GWB for large particle size MgO are shown.

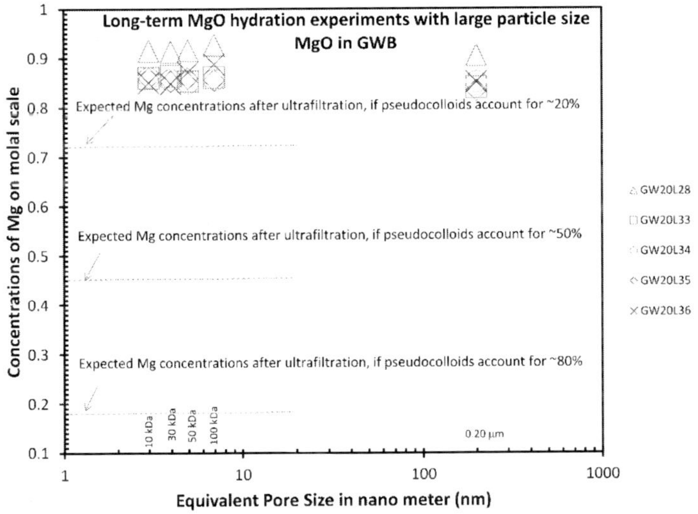

Figure 1. Magnesium concentrations in solution samples from MgO hydration experiments in GWB with large particle size MgO as a function of equivalent pore size. The experiments include GW20L28, GW20L33 through GW20L36. In each experiment, there was 3.1 g of MgO with 77 mL of GWB.

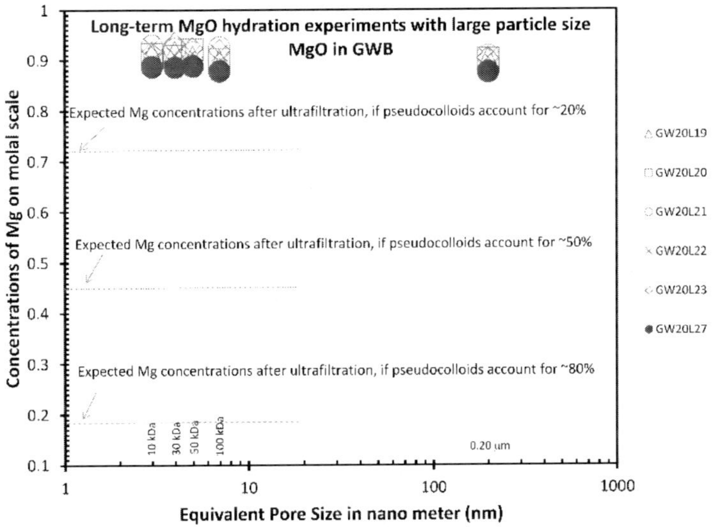

Figure 2. Magnesium concentrations in solution samples from MgO hydration experiments in GWB with large particle size MgO as a function of equivalent pore size. The experiments include GW20L19 through GW20L23, and GW20L27. In each experiment, there was 3.1 g of MgO with 77 mL of GWB.

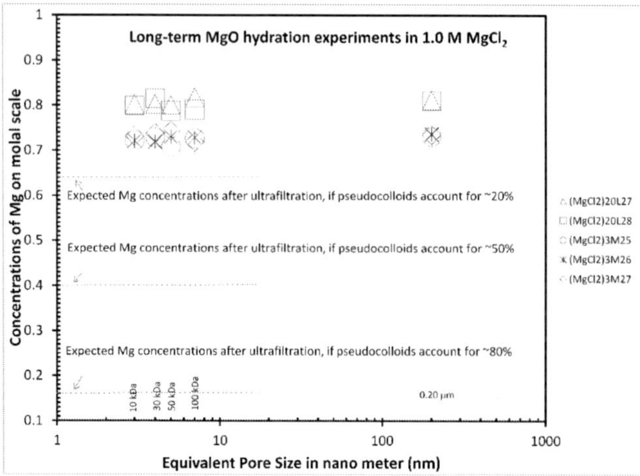

Figure 3. Magnesium concentrations in solution samples from MgO hydration experiments in a 1.0 M $MgCl_2$ solution with large particle size MgO and mixed size MgO as a function of equivalent pore size. The experiments with large particle size MgO include (MgC12) 20L27 through (MgC12)20L28. In each above experiment, there was 3.1 g of MgO with 77 mL of 1.0 M $MgCl_2$. The experiments with mixed size MgO include (MgC12)3M25 through (MgC12)3M27. In each experiment with mixed size MgO, there was 3 g of MgO with 11 mL of 1.0 M $MgCl_2$.

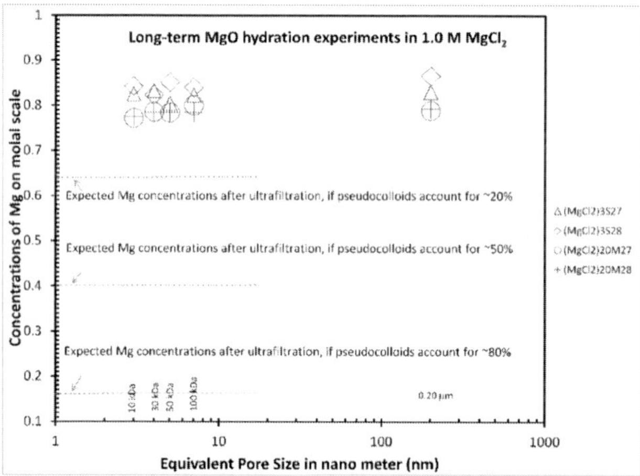

Figure 4. Magnesium concentrations in solution samples from MgO hydration experiments in a 1.0 M $MgCl_2$ solution with small particle size MgO and mixed size MgO as a function of equivalent pore size. The experiments with small particle size MgO include MgC12)3S27 through (MgC12)3S28. In each above experiment, there was 3 g of MgO with 11 mL of 1.0 M $MgCl_2$. The experiments with mixed size MgO include (MgC12)20M27 through (MgC12) 20M28. In each experiment with mixed size MgO, there was 3.1 g of MgO with 77 mL of 1.0 M $MgCl_2$.

The equivalent pore sizes used in experiments for phase separation range from 3 nm to 200 nm. Figures 1 and 2 indicate that magnesium concentrations remain constant over this range of equivalent pore sizes, suggesting the absence of mineral fragment colloids from MgO hydration products in experiments with large particle size MgO. Magnesium concentrations as a function of ultrafiltration for long-term MgO hydration experiments in 1.0 M $MgCl_2$ are presented in Figures 3 and 4. These experiments covered three particle sizes of MgO—mixed (i.e., as-received) particle size MgO, large particle size MgO and small particle size MgO. Figures 3 and 4 demonstrate that magnesium concentrations remain constant regardless of ultrafiltration, suggesting the absence of mineral fragment colloids from MgO hydration products in 1.0 M $MgCl_2$.

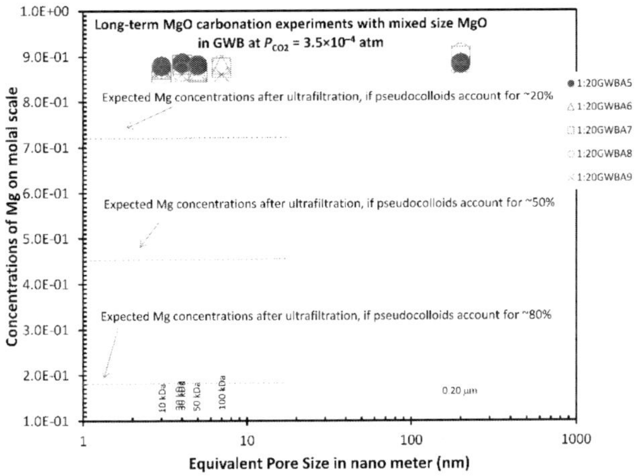

Figure 5. Magnesium concentrations in solution samples from MgO carbonation experiments in GWB with mixed particle size MgO at $P_{CO_2} = 3.5 \times 10^{-4}$ atm, as a function of equivalent pore size. The experiments include 1:20GWBA5 through 1:20GWBA9. In each experiment, there was 3.1 g of MgO with 77 mL of GWB.

In Figures 5 and 6, magnesium concentrations as a function of ultrafiltration for long-term MgO carbonation experiments in GWB are displayed. These experiments were conducted at $P_{CO_2} = 3.5 \times 10^{-4}$ atm with mixed (i.e., as-received) particle size MgO. Figures 5 and 6 indicate that magnesium concentrations remain constant regardless of ultrafiltration, suggesting the absence of mineral fragment colloids from both MgO hydration and carbonation products in GWB.

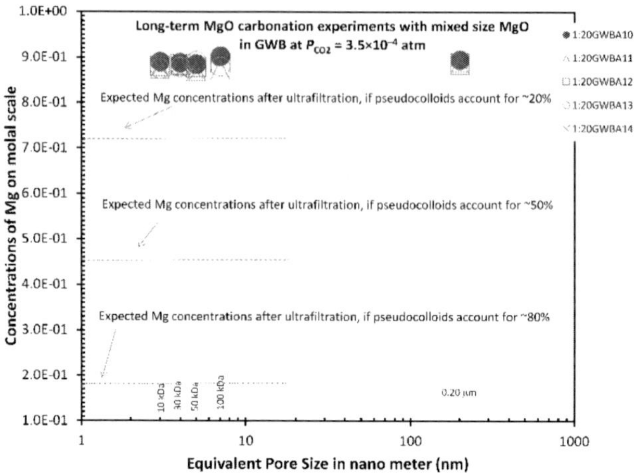

Figure 6. Magnesium concentrations in solution samples from MgO carbonation experiments in GWB with mixed particle size MgO at $P_{CO_2} = 3.5 \times 10^{-4}$ atm, as a function of equivalent pore size. The experiments include 1:20GWBA10 through 1:20GWBA14. In each experiment, there was 3.1 g of MgO with 77 mL of GWB.

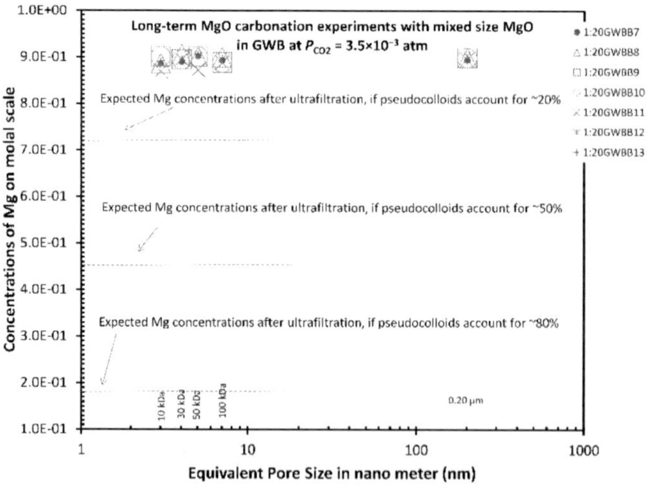

Figure 7. Magnesium concentrations in solution samples from MgO carbonation experiments in GWB with mixed particle size MgO at $P_{CO_2} = 3.5 \times 10^{-3}$ atm, as a function of equivalent pore size. The experiments include 1:20GWBB7 through 1:20GWBB13. In each experiment, there was 3.1 g of MgO with 77 mL of GWB.

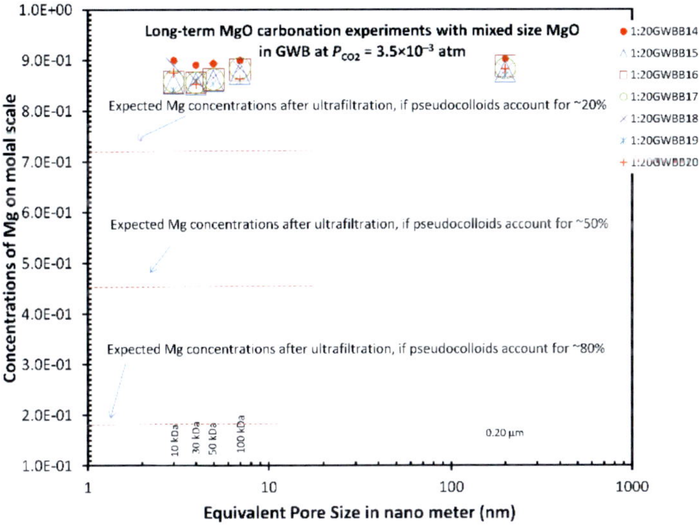

Figure 8. Magnesium concentrations in solution samples from MgO carbonation experiments in GWB with mixed particle size MgO at P_{CO_2} = 3.5 × 10^{-3} atm, as a function of equivalent pore size. The experiments include 1:20GWBB14 through 1:20GWBB20. In each experiment, there was 3.1 g of MgO with 77 mL of GWB.

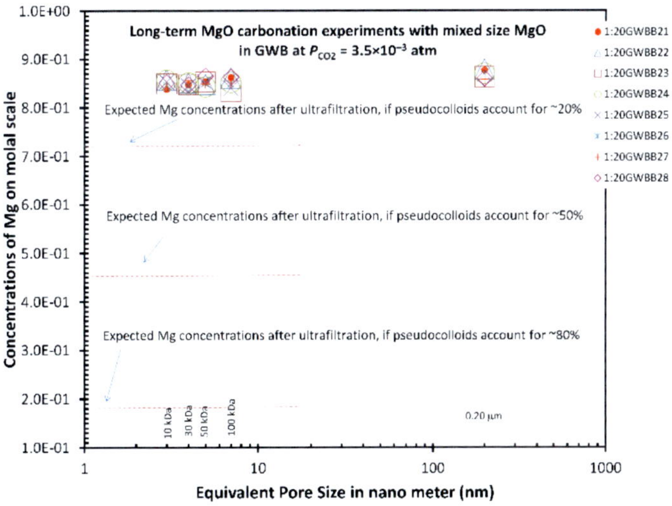

Figure 9. Magnesium concentrations in solution samples from MgO carbonation experiments in GWB with mixed particle size MgO at P_{CO_2} = 3.5 × 10^{-3} atm, as a function of equivalent pore size. The experiments include 1:20GWBB21 through 1:20GWBB28. In each experiment, there was 3.1 g of MgO with 77 mL of GWB.

In Figures 7 through 9, magnesium concentrations as a function of ultrafiltration for long-term MgO carbonation experiments in GWB are depicted. These experiments were conducted at $P_{CO_2} = 3.5 \times 10^{-3}$ atm with mixed particle size MgO. These figures indicate that magnesium concentrations remain constant regardless of ultrafiltration, suggesting the absence of mineral fragment colloids from both MgO hydration and carbonation products in GWB.

In Figures 10 through 12, the results from the carbonation experiments conducted at $P_{CO_2} = 3.5 \times 10^{-3}$ atm with large particle size MgO in GWB are shown. These figures indicate that magnesium concentrations remain constant regardless of ultrafiltration.

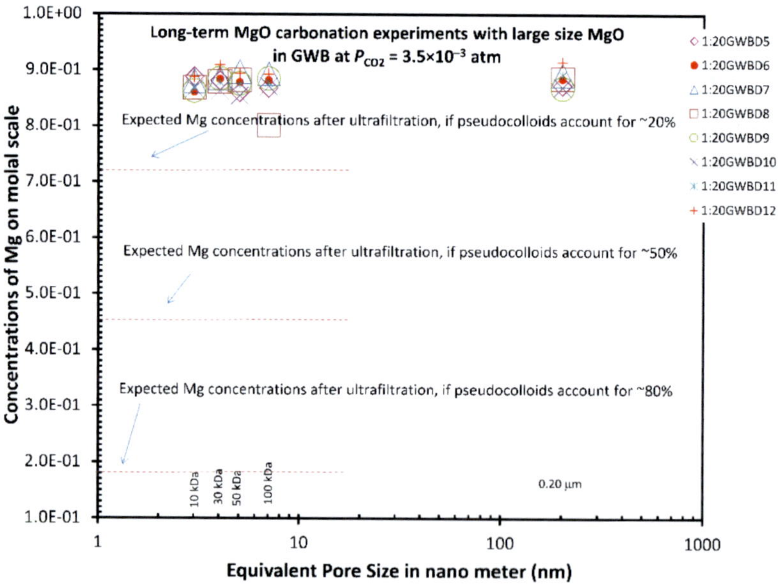

Figure 10. Magnesium concentrations in solution samples from MgO carbonation experiments in GWB with large particle size MgO at $P_{CO_2} = 3.5 \times 10^{-3}$ atm, as a function of equivalent pore size. The experiments include 1:20GWBD5 through 1:20GWBD12. In each experiment, there was 3.1 g of MgO with 77 mL of GWB.

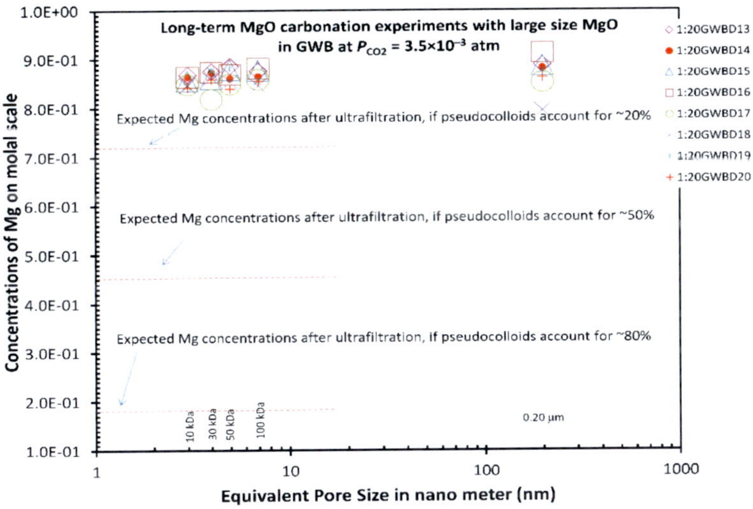

Figure 11. Magnesium concentrations in solution samples from MgO carbonation experiments in GWB with large particle size MgO at $P_{CO_2} = 3.5 \times 10^{-3}$ atm, as a function of equivalent pore size. The experiments include 1:20GWBD13 through 1:20GWBD20. In each experiment, there was 3.1 g of MgO with 77 mL of GWB.

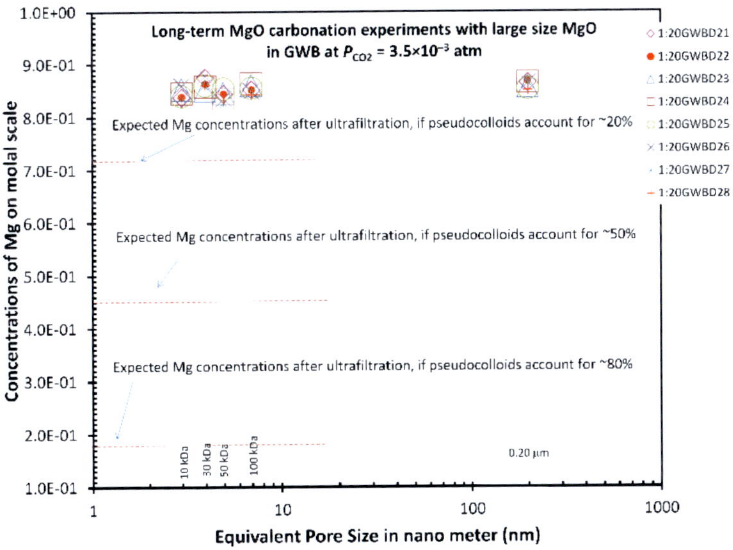

Figure 12. Magnesium concentrations in solution samples from MgO carbonation experiments in GWB with large particle size MgO at $P_{CO_2} = 3.5 \times 10^{-3}$ atm, as a function of equivalent pore size. The experiments include 1:20GWBD21 through 1:20GWBD28. In each experiment, there was 3.1 g of MgO with 77 mL of GWB.

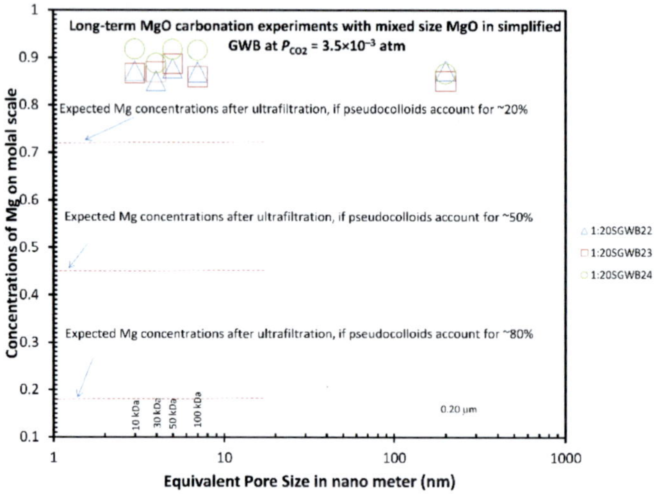

Figure 13. Magnesium concentrations in solution samples from MgO carbonation experiments in simplified GWB with mixed particle size MgO at $P_{CO_2} = 3.5 \times 10^{-3}$ atm, as a function of equivalent pore size. The experiments include 1:20SGWB22 through 1:20SGWB24. In each experiment, there was 3.1 g of MgO with 77 mL of simplified GWB.

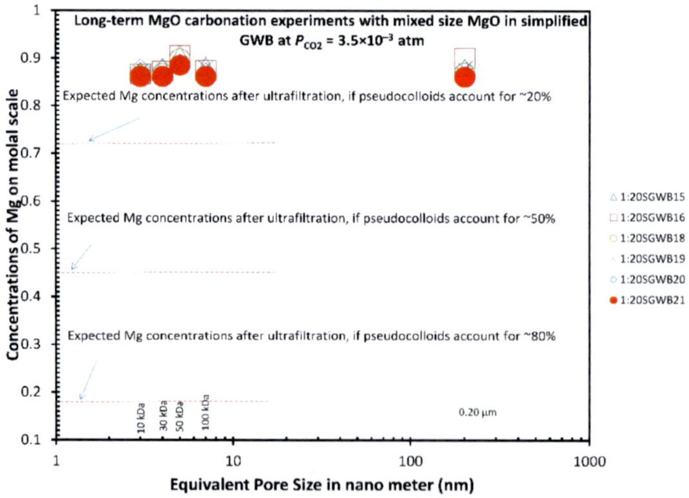

Figure 14. Magnesium concentrations in solution samples from MgO carbonation experiments in simplified GWB with mixed particle size MgO at $P_{CO_2} = 3.5 \times 10^{-3}$ atm, as a function of equivalent pore size. The experiments include 1:20SGWB15 through 1:20SGWB16, 1:20SGWB18 through 1:20SGWB21. In each experiment, there was 3.1 g of MgO with 77 mL of simplified GWB.

Figures 13 and 14 are similar to Figures 10 through 12, except that the results refer to the carbonation experiments conducted at $P_{CO_2} = 3.5 \times 10^{-3}$ atm with mixed particle size MgO in simplified GWB.

In Figure 15, the results from the carbonation experiments conducted at $P_{CO_2} = 3.5 \times 10^{-4}$ atm with prehydrated mixed particle size MgO in GWB are presented. In these experiments, MgO was prehydrated as brucite and phase 5. Therefore, these experiments actually investigated the direct carbonation of MgO from its hydration products. These experimental results indicate that magnesium concentrations remain constant regardless of ultrafiltration, suggesting the absence of mineral fragment colloids from both MgO hydration and carbonation products in GWB.

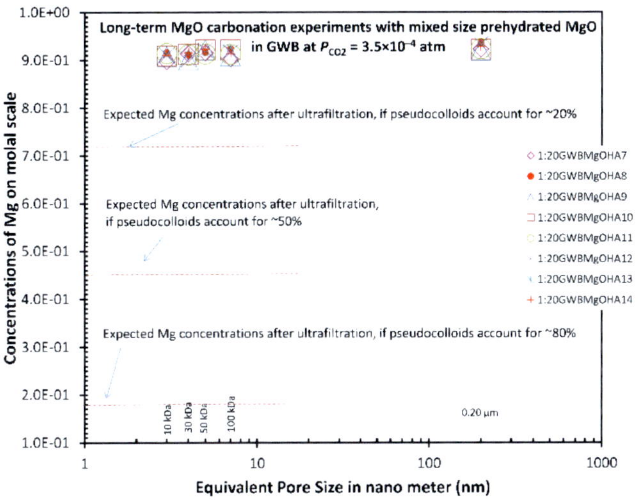

Figure 15. Magnesium concentrations in solution samples from MgO carbonation experiments in GWB with prehydrated mixed particle size MgO at $P_{CO_2} = 3.5 \times 10^{-4}$ atm, as a function of equivalent pore size. The experiments include 1:20GWBMgOHA7 through 1:20GWBMgOHA14. In each experiment, there was 3.1 g of MgO with 77 mL of GWB.

The experimental results presented in Figure 16 are similar to those in Figure 15, except that the results in Figure 16 are from the carbonation experiments conducted at $P_{CO_2} = 3.5 \times 10^{-4}$ atm with prehydrated mixed particle size MgO in simplified GWB.

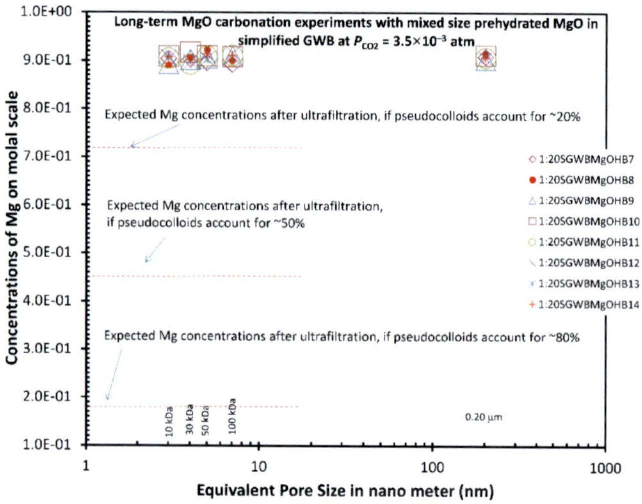

Figure 16. Magnesium concentrations in solution samples from MgO carbonation experiments in simplified GWB with prehydrated mixed particle size MgO at $P_{CO_2} = 3.5 \times 10^{-3}$ atm, as a function of equivalent pore size. The experiments include 1:20SGW BMgOHB7 through 1:20GWBMgOHB14. In each experiment, there was 3.1 g of MgO with 77 mL of simplified GWB.

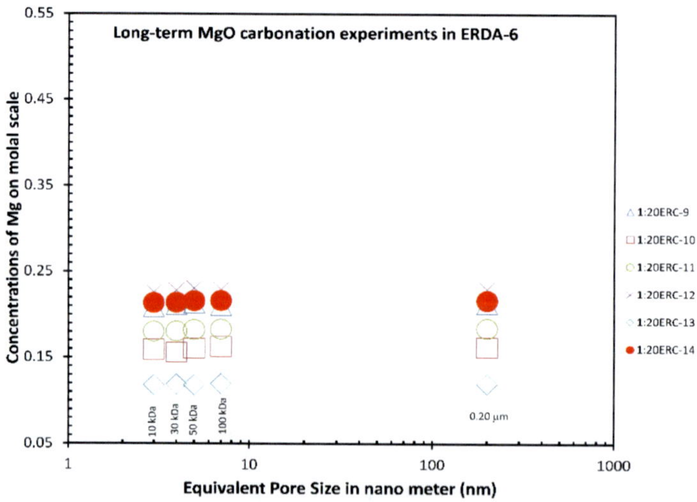

Figure 17. Magnesium concentrations in solution samples from MgO carbonation experiments in ERDA-6 with small particle size MgO at $P_{CO_2} = 3.5 \times 10^{-3}$ atm, as a function of equivalent pore size. The experiments include 1:20ERC-9 through 1:20ERC-14. In each experiment, there was 3.1 g of MgO with 77 mL of ERDA-6.

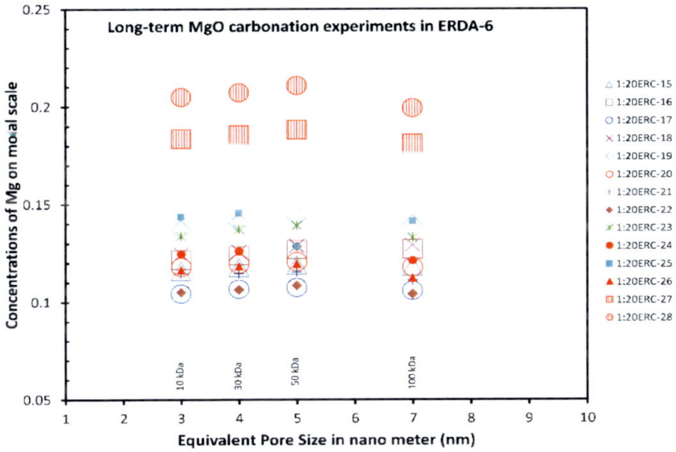

Figure 18. Magnesium concentrations in solution samples from MgO carbonation experiments in ERDA-6 with small particle size MgO at $P_{CO_2} = 3.5 \times 10^{-3}$ atm, as a function of equivalent pore size. The experiments include 1:20ERC-15 through 1:20ERC-28. In each experiment, there was 3.1 g of MgO with 77 mL of ERDA-6.

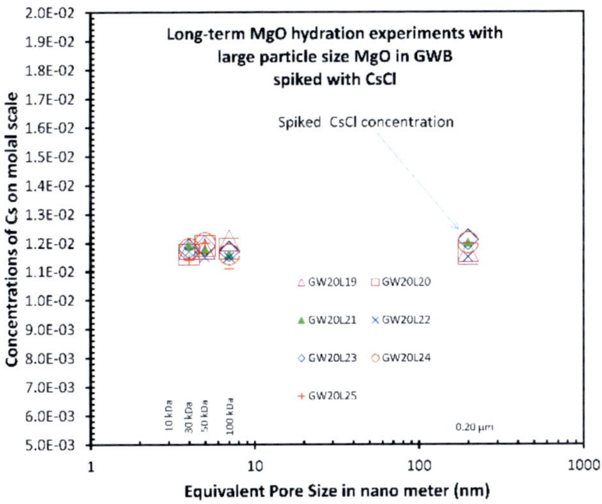

Figure 19. Cesium concentrations in solution samples from MgO hydration experiments in GWB with large particle size MgO as a function of equivalent pore size. The experiments include GW20L19 through GW20L25. In each experiment, there was 3.1 g of MgO with 77 mL of GWB.

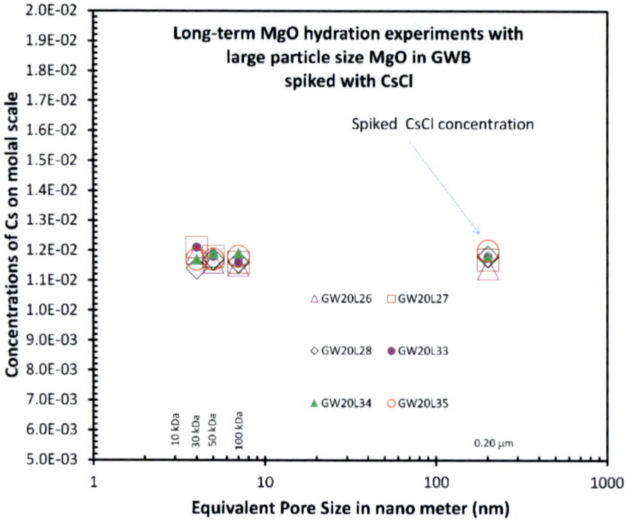

Figure 20. Cesium concentrations in solution samples from MgO hydration experiments in GWB with large particle size MgO as a function of equivalent pore size. The experiments include GW20L26 through GW20L28, GW20L33 through GW20l35. In each experiment, there was 3.1 g of MgO with 77 mL of GWB.

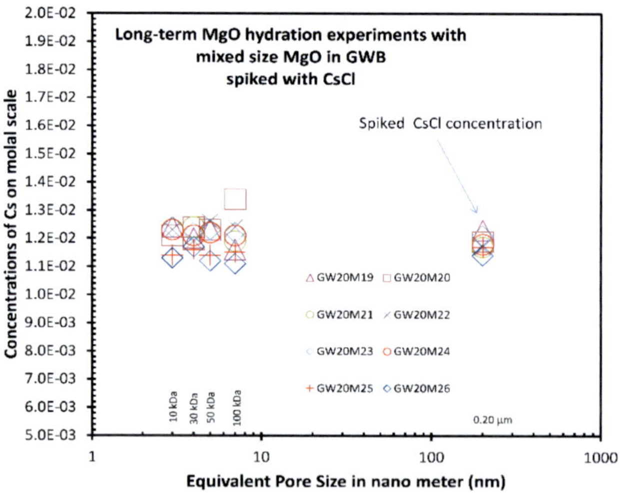

Figure 21. Cesium concentrations in solution samples from MgO hydration experiments in GWB with mixed particle size MgO as a function of equivalent pore size. The experiments include GW20M19 through GW20M26. In each experiment, there was 3.1 g of MgO with 77 mL of GWB.

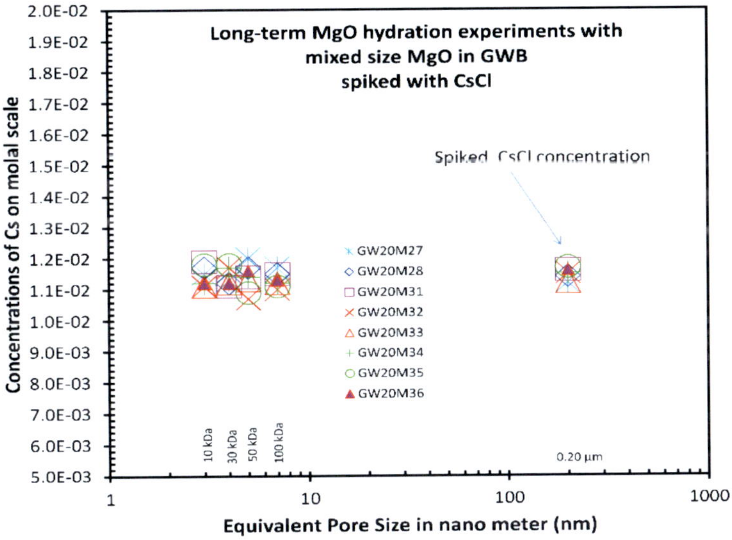

Figure 22. Cesium concentrations in solution samples from MgO hydration experiments in GWB with mixed particle size MgO as a function of equivalent pore size. The experiments include GW20M27 through GW20M28, GW20M31 through GW20M36. In each experiment, there was 3.1 g of MgO with 77 mL of GWB.

In Figures 17 and 18, magnesium concentrations as a function of ultrafiltration for long-term MgO carbonation experiments in ERDA-6 are displayed. These experiments were conducted at $P_{CO_2} = 3.5 \times 10^{-3}$ atm with small particle size MgO. Figures 17 and 18 indicate that magnesium concentrations remain constant regardless of ultrafiltration, suggesting the absence of mineral fragment colloids from both MgO hydration and carbonation products in EDRA-6.

Cesium Concentrations as a Function of Equivalent Pore Sizes

In the following, cesium concentrations as a function of equivalent pore size for samples taken from the SNL long-term MgO hydration and carbonation experiments in various solutions are presented.

In Figures 19 and 20, cesium concentrations as a function of ultrafiltration for long-term MgO hydration experiments in GWB are displayed. These experiments were conducted with large particle size MgO. Cesium concentrations are constant regardless of equivalent pore sizes, indicating the absence of mineral fragment colloids.

The results presented in Figures 21 and 22 are similar to those in Figures 19 and 20. The experiments presented in Figures 21 and 22 were conducted in GWB with mixed particle size MgO.

In Figure 23, the results from the long-term MgO hydration experiments in 1.0 $MgCl_2$ are presented. These experiments were conducted with large particle size MgO. The results are similar to those in Figures 19 and 20 in GWB.

In Figures 24 and 25, the results from the carbonation experiments conducted at $P_{CO_2} = 3.5 \times 10^{-4}$ atm with mixed particle size MgO in GWB are depicted. There is no dependence on equivalent pore size for cesium concentrations.

The results in Figures 26 through 28 are similar to those in Figures 24 and 25, except that the results in Figures 26 through 28 are from the carbonation experiments conducted at $P_{CO_2} = 3.5 \times 10^{-3}$ atm, which is one order of magnitude higher than that in Figures 24 and 25.

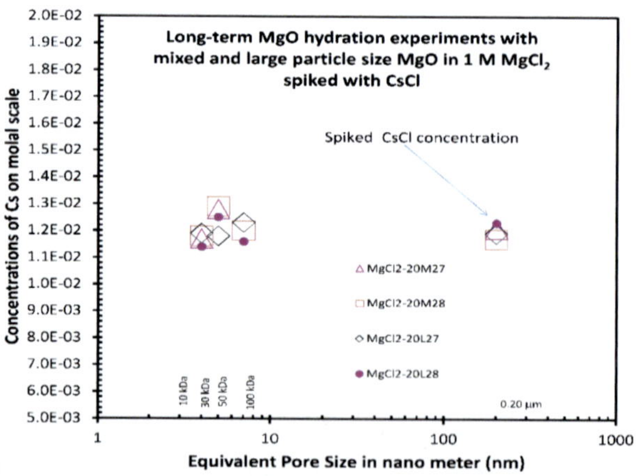

Figure 23. Cesium concentrations in solution samples from MgO hydration experiments in 1.0 M $MgCl_2$ with mixed and large particle size MgO as a function of equivalent pore size. The experiments with mixed particle size MgO include MgCl2-20M27 and MgCl2-20M28. The experiments with large particle size MgO include MgCl2-20L27 and MgCl2-20L28. In each experiment, there was 3.1 g of MgO with 77 mL of 1.0 M $MgCl_2$.

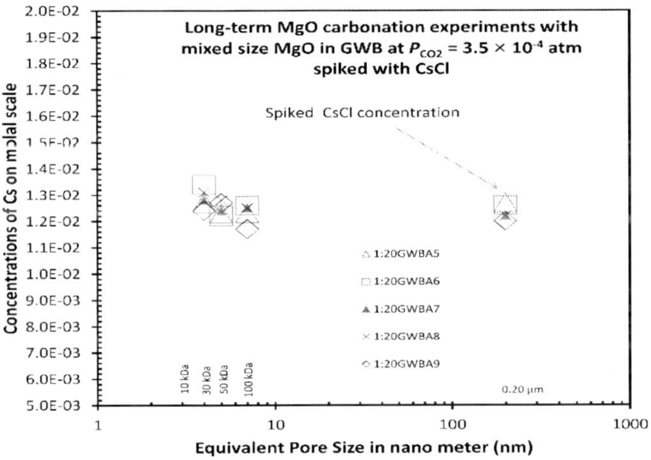

Figure 24. Cesium concentrations in solution samples from MgO carbonation experiments in GWB with mixed particle size MgO at P_{CO_2} = 3.5 × 10^{-4} atm, as a function of equivalent pore size. The experiments include 1:20GWBA5 through 1:20GWBA9. In each experiment, there was 3.1 g of MgO with 77 mL of GWB.

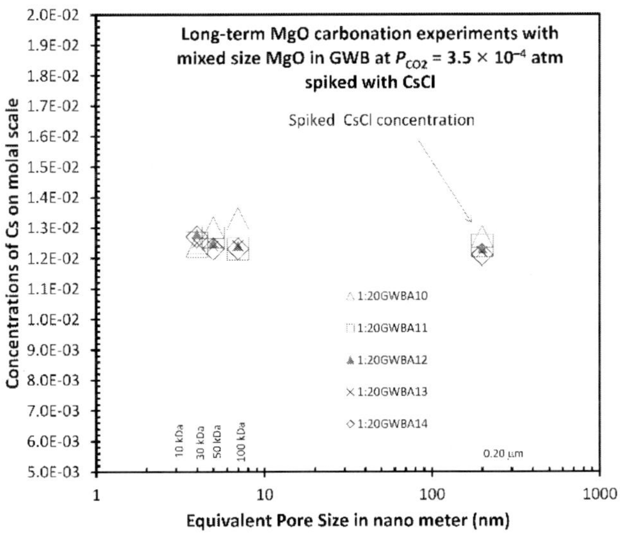

Figure 25. Cesium concentrations in solution samples from MgO carbonation experiments in GWB with mixed particle size MgO at P_{CO_2} = 3.5 × 10^{-4} atm, as a function of equivalent pore size. The experiments include 1:20GWBA10 through 1:20GWBA14. In each experiment, there was 3.1 g of MgO with 77 mL of GWB.

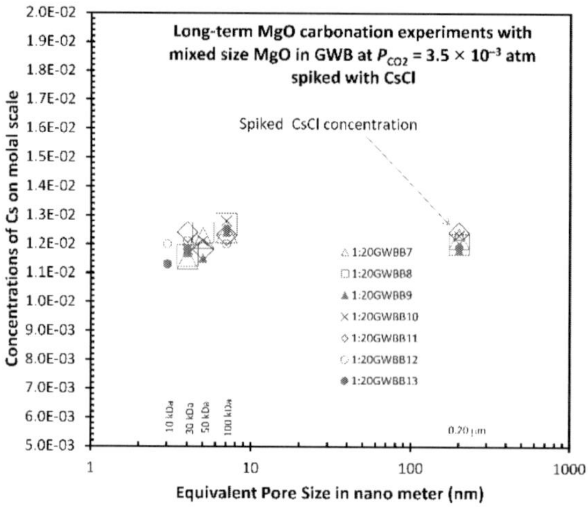

Figure 26. Cesium concentrations in solution samples from MgO carbonation experiments in GWB with mixed particle size MgO at $P_{CO_2} = 3.5 \times 10^{-3}$ atm, as a function of equivalent pore size. The experiments include 1:20GWBB7 through 1:20GWBB13. In each experiment, there was 3.1 g of MgO with 77 mL of GWB.

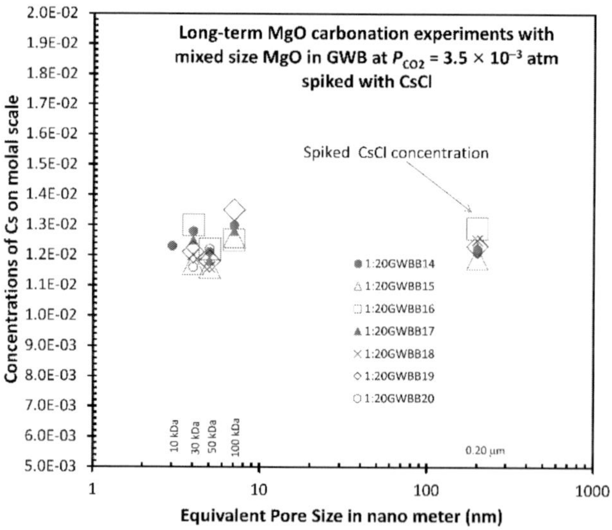

Figure 27. Cesium concentrations in solution samples from MgO carbonation experiments in GWB with mixed particle size MgO at $P_{CO_2} = 3.5 \times 10^{-3}$ atm, as a function of equivalent pore size. The experiments include 1:20GWBB14 through 1:20GWBB20. In each experiment, there was 3.1 g of MgO with 77 mL of GWB.

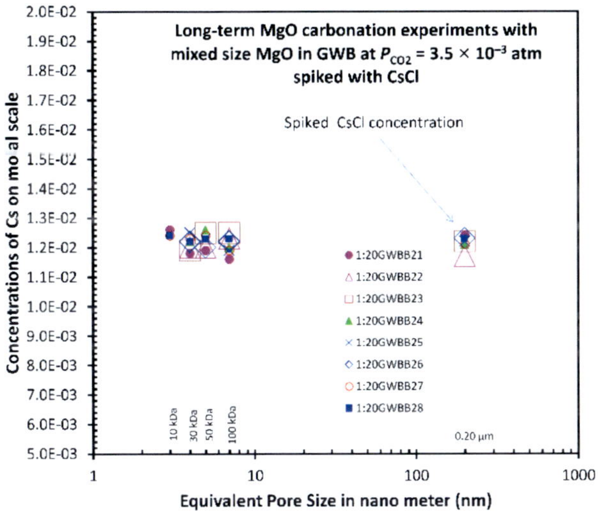

Figure 28. Cesium concentrations in solution samples from MgO carbonation experiments in GWB with mixed particle size MgO at $P_{CO_2} = 3.5 \times 10^{-3}$ atm, as a function of equivalent pore size. The experiments include 1:20GWBB21 through 1:20GWBB28. In each experiment, there was 3.1 g of MgO with 77 mL of GWB.

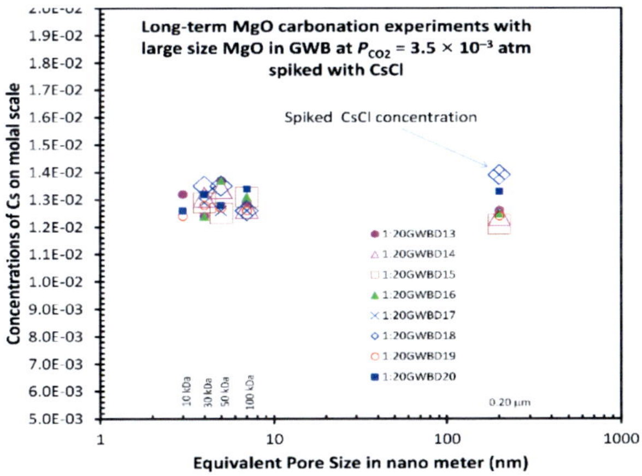

Figure 29. Cesium concentrations in solution samples from MgO carbonation experiments in GWB with large particle size MgO at $P_{CO_2} = 3.5 \times 10^{-3}$ atm, as a function of equivalent pore size. The experiments include 1:20GWBD13 through 1:20GWBD20. In each experiment, there was 3.1 g of MgO with 77 mL of GWB.

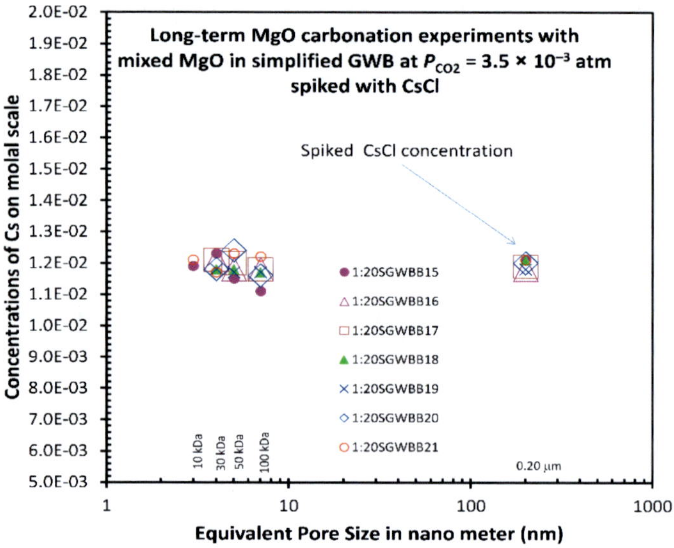

Figure 30. Cesium concentrations in solution samples from MgO carbonation experiments in simplified GWB with mixed particle size MgO at $P_{CO_2} = 3.5 \times 10^{-3}$ atm, as a function of equivalent pore size. The experiments include 1:20SGWBB15 through 1:20SGWBB21. In each experiment, there was 3.1 g of MgO with 77 mL of simplified GWB.

The results presented in Figure 29 are similar to those in Figures 26 through 28. The experiments in Figure 29 were with large particle size MgO, whereas those in Figures 26 through 28 were with mixed particle size MgO.

The results presented in Figure 30 are similar to those in Figure 29. The experiments in Figure 29 were with large particle size MgO in GWB, whereas those in Figure 30 were with mixed particle size MgO in simplified GWB.

In Figure 31, the results from the carbonation experiments conducted at $P_{CO_2} = 3.5 \times 10^{-3}$ atm with prehydrated large particle size MgO in GWB are depicted. As mentioned previously, MgO was prehydrated as brucite and phase 5. Therefore, these experiments investigated the direct carbonation of hydration products. These experimental results indicate that cesium concentrations remain constant regardless of ultrafiltration, suggesting the absence of mineral fragment colloids from both MgO hydration and carbonation products in GWB.

The results in Figure 32 are similar to those in Figure 31, except for that the carbonation experiments with prehydrated MgO in Figure 32 were conducted in simplified GWB.

In summary, the experimental results demonstrate that magnesium and cesium concentrations are independent of ultrafiltration, indicating the absence of mineral colloids associated with MgO hydration and carbonation processes. These experiments covered a wide range of conditions. In terms of MgO particle sizes, they included mixed (i.e., as-received) particle size MgO, large particle size MgO (i.e., 1 mm—2 mm), and small particle size MgO (i.e., < 75 μm). Regarding the solution medium, they included 1.0 $MgCl_2$, simplified GWB (1.0 M $MgCl_2$ + 3.6 M NaCl), GWB, and ERDA-6. With regard to the partial pressure of CO_2(g), they included $P_{CO_2} = 3.5 \times 10^{-4}$ atm and $P_{CO_2} = 3.5 \times 10^{-3}$ atm.

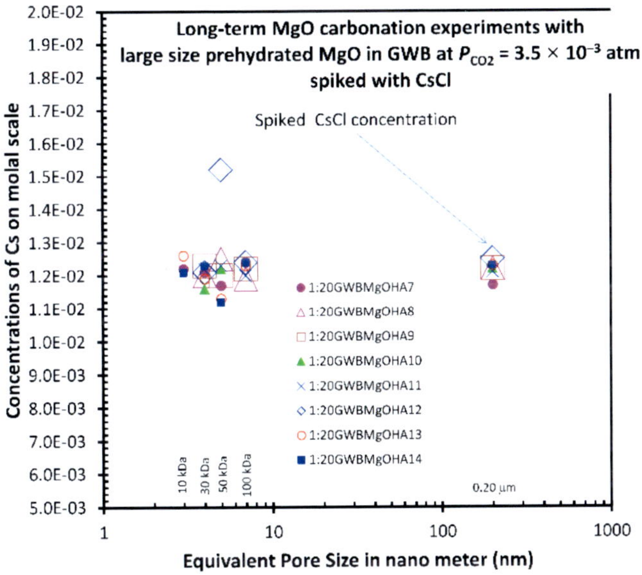

Figure 31. Cesium concentrations in solution samples from MgO carbonation experiments in GWB with prehydrated large particle size MgO at $P_{CO_2} = 3.5 \times 10^{-3}$ atm, as a function of equivalent pore size. The experiments include 1:20GWBMgOHA7 through 1:20GWBMgOHA13. In each experiment, there was 3.1 g of MgO with 77 mL of GWB.

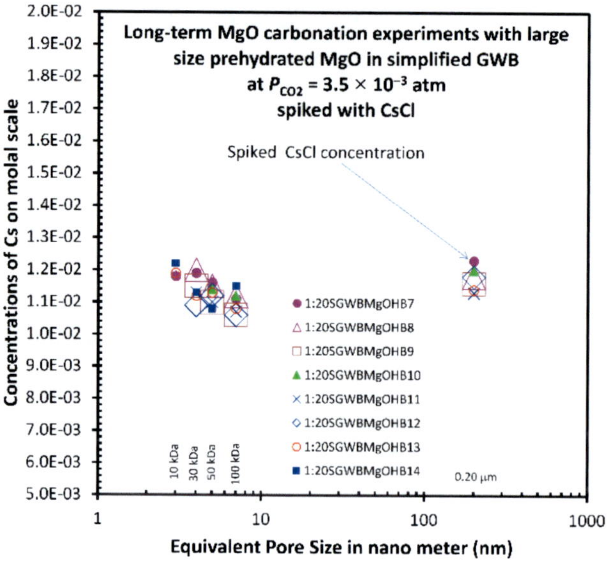

Figure 32. Cesium concentrations in solution samples from MgO carbonation experiments in simplified GWB with prehydrated large particle size MgO at $P_{CO_2} = 3.5 \times 10^{-3}$ atm, as a function of equivalent pore size. The experiments include 1:20SGWBMgOHB7 through 1:20SGWBMgOHB14. In each experiment, there was 3.1 g of MgO with 77 mL of simplified GWB.

CONCLUSION

This work clearly demonstrates the absence of mineral fragment colloids in the SNL long-term MgO hydration and carbonation. This conclusion is reached based on two independent experimental approaches. The absence of mineral fragment colloids in the WIPP brines is due to the chemistry of the WIPP brines. The WIPP brines are characterized by high ionic strengths with GWB and ERDA-6 having ionic strengths of 8.26 mol•kg^{-1} and 5.82 mol•kg^{-1}, respectively (Xiong and Lord, 2008). Furthermore, these brines have high concentrations of both monovalent and divalent ions. The ERDA-6 brine has 5.35 mol•kg^{-1}, 0.106 mol•kg^{-1}, 0.0209 mol•kg^{-1}, 0.0132 mol•kg^{-1} for Na^+, K^+, Mg^{2+}, and Ca^{2+} (Xiong and Lord, 2008), respectively. GWB has 4.04 mol•kg^{-1}, 0.532 mol•kg^{-1}, 1.16 mol•kg^{-1}, 0.0163 mol•kg^{-1} for Na^+, K^+, Mg^{2+}, and Ca^{2+} (Xiong and Lord,

2008), respectively. These concentrations are much higher than the critical coagulation concentrations. The direct consequence of the chemistry of the WIPP brines is that mineral fragment colloids that might be derived from MgO are unstable in these brines. The absence of mineral fragment colloids, as demonstrated experimentally by this study, obviously illustrates the role of the WIPP brine chemistry in destabilization of mineral fragment colloids.

The absence of MgO hydration and carbonation product mineral fragment colloids in the brines associated with salt formations, as indicated by this work, has important implications to disposal of nuclear waste in salt formations. The high ionic strength brines associated with salt formations provide favorable near-field geochemical conditions that will de-stabilize mineral fragment colloids. The absence of colloids of any kind is a favorable result for the performance of a nuclear waste repository in salt formations.

ACKNOWLEDGMENTS

Sandia National Laboratories is a multimission laboratory managed and operated by National Technology and Engineering Solutions of Sandia, LLC., a wholly owned subsidiary of Honeywell International, Inc., for the U.S. Department of Energy's National Nuclear Security Administration under contract DE-NA-0003525. This research is funded by the WIPP programs administered by the Office of Environmental Management (EM) of the U.S. Department of Energy. SAND2017-6865B.

REFERENCES

Allen, L.H. and E. Matijević, 1969. "Stability of colloidal silica: I. Effect of simple electrolytes." *Journal of Colloid and Interface Science* 31:287–296.

Allen, L.H. and E. Matijevic, 1971. "Stability of colloidal silica: II. Ion exchange." *Journal of Colloid and Interface Science* 33:420–429.

Altmaier, M., V. Neck, and Th. Fanghanel, 2004. "Solubility and colloid formation of Th(IV) in concentrated NaCl and $MgCl_2$ solution." *Radiochimica Acta* 92:537–543.

Apalkov, Gleb, Alexander Dyachenko, Andrey Zhabin, and Sergey Smirnov, 2016. "Development of a purification technology for treatment of medium-and low-activity radioactive waste of radiochemical production from Co-60 and Cs-137." *MATEC Web of Conferences* 85:01021. EDP Sciences.

Bascetin, E., H. Haznedaroglu, and A.Y. Erkol, 2003. "The adsorption behavior of cesium on silica gel." *Applied Radiation and Isotopes 59*:5–9.

Bitea, C., C. Walther, J.-I Kim, H. Geckeis, T. Rabung, F.J. Scherbaum, D.G. Cacuci, 2003. "Time-resolved observation of ZrO_2-colloid agglomeration." *Colloids and Surfaces A: Physicochem. Eng. Aspects* 215:55–66.

Chen, G., M, Flury, J.B. Harsh, and P.C. Lichtner, 2005. "Colloid-facilitated transport of cesium in variably saturated Hanford sediments." *Environmental Science & Technology* 39:3435–3442.

Cheng, Tao, and James E. Saiers, 2010. "Colloid-facilitated transport of cesium in vadose-zone sediments: the importance of flow transients." *Environmental Science & Technology* 44:7443–7449.

Cheng, Tao, and James E. Saiers. 2015. "Effects of dissolved organic matter on the co-transport of mineral colloids and sorptive contaminants." *Journal of Contaminant Hydrology* 177:148–157.

Deng, H.-R., Nemer, M., and Xiong, Y.-L., 2006. *"Test Plan TP 06-03 Experimental Study of MgO Reaction Pathways and Kinetics, Revision 1."* Sandia National Laboratories, Carlsbad, NM.

Deng, Haoran, Yongliang Xiong, Martin Nemer, and Shelly Johnsen, 2008. "Experimental Work Conducted on MgO Characterization and Hydration." *MRS Proceedings*1124: 1124-Q05. Cambridge University Press.

Deng, H.-R., Y.-L. Xiong, M. Nemer, and S. Johnsen, 2009. Experimental work conducted on MgO long-term hydration, *2008 Milestone Report*. Sandia National Laboratories, Carlsbad, NM, ERMS 551421.

Itagaki, H., S. Tanaka, M. Yamawaki, 1991. "Neptunium chemical behavior in underground environments using ultrafiltration and centrifugation." *Radiochimica Acta* 52/53:91–94.

Kelly, J.W., R. Aguilar, and H.W. Papenguth, 1996. *Contribution of mineral-fragment type pseudo-colloids to the mobile actinide source term of the Waste Isolation Pilot Plant (WIPP)*. Sandia National Laboratories, Albuquerque, NM, SAND96-2742J.

Kim, Intae, and Guebuem Kim, 2015. "Role of colloids in the discharge of trace elements and rare earth elements from coastal groundwater to the ocean." *Marine Chemistry* 176:126–132.

Kim, J.I., B. Delakowitz, P. Zeh, D. Klotz, D. Lazik, 1994. "A column experiment for the study of colloidal radionuclide migration in Gorleben aquifer systems." *Radiochimica Acta* 66/67:165–171.

Olofsson, U., and B. Allard, 1986. "Formation and transport of americium pseudocolloids in aqueous systems." *SKBF/KBS Technical Report -TR 86-02*, Swedish Nucl. Fuel and Waste Management Co., Stockholm.

Olofsson, U., B. Allard, M. Bengtsson, B. Torstenfelt, K. Andersson, 1983. "Formation and Properties of Actinide Colloids." *SKBF/KBS Technical Report -TR 83-08*, Swedish Nucl. Fuel and Waste Management Co., Stockholm.

Ross, J.M. and R.M. Sherrell, 1999. "The role of colloids in trace metal transport and adsorption behavior in New Jersey Pinelands streams." *Limnol. Oceanogr* 44:1019–1034.

Sasaki, Takayuki, Yuu Takeno, Akira Kirishima, and Nobuaki Sato, 2015. "Leaching test of gamma-emitting Cs, Ru, Zr, and U from neutron-irradiated UO_2/ZrO_2 solid solutions in non-filtered surface seawater: Fukushima NPP Accident Related." *Journal of Nuclear Science and Technology* 52:147–151.

Schuessler, W., B. Kienzler, S. Wilhelm, V. Neck, and J.I. Kim, 2001. "Modeling of Near Field Actinide Concentrations in Radioactive

Waste Repositories in Salt Formations: Effect of Buffer Materials." *Mat. Res. Soc. Symp. Proc.* 663:791.

Silva, R.J., H. Nitsche, 1995. "Actinide environmental chemistry." *Radiochimica Acta* 70/71:377–396.

U.S. National Academy of Sciences Committee on Waste Disposal. 1957. *The Disposal of Radioactive Waste on Land.* Publication 519. Washington, DC: National Academy of Sciences–National Research Council.

van Olphen, H. 1991. *An Introduction to Clay Colloid Chemistry: For Clay Technologists, Geologists, and Soil Scientists.* 2nd ed. Malabar, FL: Krieger Pub. Co. (Reprint of 1977- 2nd edition originally published by John Wiley& Sons.)

Wang, Dengjun, Yan Jin, and Deb P. Jaisi, 2015. "Cotransport of hydroxyapatite nanoparticles and hematite colloids in saturated porous media: Mechanistic insights from mathematical modeling and phosphate oxygen isotope fractionation." *Journal of Contaminant Hydrology* 182:194–209.

Wen, L.S., Santschi, P.H., Paternostro, C. and Gill, G., 1999. "Estuarine trace metal distributions in Galveston Bay I: importance of colloidal forms in the speciation of the dissolved phase." *Mar. Chem* 63:185–212.

Xiong, Y.-L., 2008. *SP 20-4, Preparing Synthetic Brines for Geochemical Experiments, Revision 2*. Sandia National Laboratories, Carlsbad, NM.

Xiong, Y.-L., and A.C.S. Lord, 2008. "Experimental investigations of the reaction path in the MgO–CO_2–H_2O system in solutions with ionic strengths, and their applications to nuclear waste isolation." *Applied Geochemistry* 23:1634–1659.

Xiong, Y.-L., L.H. Brush, J. Garner, and J.J Long, 2010a. *Responses to three EPA comments pertaining to comparisons of measured and predicted dissolved and colloidal Th(IV) and Am(III) concentrations, Revision 1*. Supersedes ERMS 553409. Sandia National Laboratories, Carlsbad, NM, ERMS 553595

Xiong, Y.-L., H. Deng, M. Nemer, S. Johnsen, 2010b. "Experimental determination of the solubility constant for magnesium chloride hydroxide hydrate ($Mg_3Cl(OH)_5 \cdot 4H_2O$, phase 5) at room temperature, and its importance to nuclear waste isolation in geological repositories in salt formations." *Geochimica et Cosmochimica Acta* 74:4605–4611.

Zhao, P.-H., S.S. Steward, 1997. *Literature review of intrinsic colloids related to spent fuel waste package release rates*. Lawrence Livermore National Laboratory, UCRL-ID-126039.

In: Solution Chemistry
Editor: Yongliang Xiong
ISBN: 978-1-53613-101-7

Chapter 5

CONFORMATIONAL RESTRUCTURATION OF IONIC POLYSULFONES IN SOLUTION: PERSPECTIVES TOWARDS SMART BIOMATERIALS

Anca Filimon*
Department of Polymer Physical Chemistry,
"Petru Poni" Institute of Macromolecular Chemistry, Iasi, Romania

ABSTRACT

New functionalized polysulfones are emerging from the academic laboratory into the biomedical field, providing new functionality for the achievement of performant bioactive membranes. The present chapter records recent progress in the development, investigation, and possible application of polysulfonic films able to form various microstructures and analyzes why some properties of modified polysulfones, particularly the viscometric and rheological ones, should be considered for their extended biomedical use. So, extends formulation of thermodynamic description into a multidomain theory, which permits a simultaneous investigation of molecular, physical, and chemical properties. This option appears

* Corresponding Author Email: capataanca@yahoo.com.

particularly helpful to forecast the behavior of joint solutions of molecular linear chains or charged polymers. Finally, the recent advances made for improving the performance of these polymers in biomedical applications will be highlighted.

Keywords: functionalized polysulfones, solution properties, polyelectrolyte effect, specific interactions, biomedical applications

INTRODUCTION

In recent years, there has been an increasing need for sustainable medical therapeutics in order to treat diseases that compromise normal functions of the human body, or even for aesthetic purposes. The fundamental concept underlying the sustainable medical treatment relies on research of polymeric materials with unique physicochemical properties and high potential to use in obtaining valuable products, that is, biocompatible and biodegradable materials. Therefore, these smart materials involving the development of medical devices, traditional implants and prostheses - to include tissue engineering and control drug delivery vehicles-, are aimed for numerous applications, such as information processing, storage and retrieval, as well as the molecular recognition similar to the biological systems. In this context, biomaterials play a important role because the interaction between cells and biomaterials determine the success or failure of most tissue engineering approaches (Bettinger 2011; Filimon et al., 2015). Additionally, in most applications, a biomaterial, once designed, has to be stable and consequently, has to maintain its structure and morphology under various environmental conditions.

According to above-mentioned, depending on their origin (as either natural or synthetic), polymers may to produce biomaterials with wide-ranging and reproducible properties by tailored variations of the components and synthetic processes. Among the synthetic and biocompatible polymeric materials available and suitable for biomedical applications, polysulfones (polysulfones, poly(ethersulfone)s, or poly(phenyl sulfone)s) represent one such class that currently has attracting attention. It is known that polysulfones (PSFs), due to superior properties, namely good optical characteristics, high thermal and chemical stabilities, mechanical strength, resistance to extreme pH values (Barikani and Shahram 2000; Higuchi et al., 1992), as well as the film-forming properties, their can be targeted to a particular application. Owing to the high-performance properties, PSFs are used in materials science, biology, and polymer science, being explored for numerous applications, such as filtration membranes (Du and Zhao 2004a; 2004b; Liu et al., 2011; Nady et al., 2011), coatings (Sivaraman et al., 2012), composites (Sanchez et al., 2007; Filimon et al., 2014; 2016), microelectronic devices (von Kraemer et al., 2006), thin-film technology (Kuroiwa 1995), biomaterials (Filimon et al., 2009; 2015), and fuel cells (Karlsson and Jannasch 2004; Manea and Mulder 2002). Thereby, desirable combination between polysulfones' properties and their availability makes them attractive materials for the membranes manufacture used in biomedical applications.

Synthetic effort in designing polymers carrying functional groups in the main or side chain, for providing superior properties, has been reported in recent years (Chandima and Jagath 2007). In this context, ionic polysulfones containing quaternary ammonium side groups have quite different properties than those with PSFs and, from a practical viewpoint, some distinct features, such as solubility (Barikani and Shahram 2000; Ioan et al., 2006; Filimon et al., 2007) hydrophilicity, and high water permeability - property with special interest for biomedical applications - (Higuchi et al., 1992), biocompatibility and antimicrobial activity (Dizman et al., 2014; Filimon et al., 2009; 2015), and better separation (Du and Zhao 2004b; Liu et al., 2011). Additionally, a proven tool to generate new stimuli-sensitive materials with tunable, reproducible, and mathematically predictable physical properties is the use the polymers blending concept. Therefore, composites based on functionalized polysulfones, due to their extraordinary characteristics and high degree of tailorability, can meet the requirements for applications as biomaterials. Consequently, it is of interest to find out whether functionalized PSFs can be designed for specific applications, in blends with synthetic polymers and to establish their impact on different properties.

Biocompatibility, bioadhesivity, and the ability of the positively charged side chain (controlled by thermodynamic quality of the solvent, ionic strength) to interact with other macromolecules or neutral compounds with low molecular weight assure an efficient application of functionalized PSFs in different domains. Considering the vast possibilities to use the new materials in biomedicine, this chapter seeks to consolidate all of the information concerning the implications of the classical and modern ways in designing the new active systems in accordance with the phenomena underlying to tailor their biocompatibility. Results would provide an insight into the development of high-performance membranes with enhanced biocompatibility and bacterial resistance by means of antibacterial agents.

Polyelectrolyte Effect of Quaternized Polysulfones on the Chain Conformation

In polymer science and engineering, the study of polymers/ polyelectrolytes in solution, as well as knowledge of the specific interactions developed in multicomponent systems represent a fundamental step in processing of advanced materials. Progress recorded in polymer chemistry have allowed designing of the ionic polymers with specific properties and also, the anticipation of their complex conformations, due to electrostatic interactions and rearrangement of counterions in solution, by the control of their physicochemical properties in solution.

As generally known, viscosity studies provide a reasonable evaluation of the conformational behavior of complex polymer structures in solution. The viscometric data are related to the structural characteristics of the polycation, namely size and charge density, but also to environmental properties, such as ionic strength, pH, and addition of other solvents or salts (Kotlyarevskaya et al., 2005).

Theoretical evaluation of polyelectrolyte solution's viscosity it remains among the least well understood systems due to its complicated behavior in solution, as a result of the combined action of electrostatic and hydrodynamic interactions. High complexity of these systems is generated by the cumulative effects derived from long-chain molecules with those derived from charge interactions. The study of polyelectrolyte solutions differs considerably from those of uncharged polymers. This statement arises from the long-range Coulomb interactions between charges which lead to a variety of effects (Dobrynin and Rubinstein 2005). As a consequence, the scaling laws, so useful for the description of uncharged systems, are difficult to apply for polyelectrolyte.

Recent studies (Lu et al., 2013; Filimon et al., 2014b) have showed that polymer mixtures in solution often exhibit properties superior to those of each individual polymer component. These studies contribute to the knowledge on the nature of the interactions manifested in the multicomponent systems and by synergistic effects, such combinations may possess unique properties different from those of the pure components.

Understanding the properties of any material/composite in depth is the first step towards elucidating its potential for possible applications. Literature (Ioan et al., 2006a; 2006b; Filimon et al., 2007; 2010) has pointed out the outstanding role polyelectrolytes play in the field of life science. From this reason, the following sections cover the aspects on the properties of some new functionalized PSFs - synthesized by quaternization of chloromethylated polysulfones with N,N-dimethylethanolamine (DMEtA), N,N-dimethylethylamine (DMEA), N,N-dimethylbutylamine (DMBA), and N,N-dimethyloctylamine (DMOA) (Avram 2001; Avram and Filimon 2015), and also of biomaterial composites (functionalized PSFs blends with polyvinyl alcohol (PVA), and/or cellulose acetate phthalate (CAP), polystyrene (PS) and/or poly(4-vinylpyridine) (P4VP)).

Influence of Charge Density on the Hydrodynamic Volume in Polyelectrolyte Systems

In the field literature (de Gennes et al., 1976; Odijk 1979; Dobrynin et al., 1995; Muthukumar 1997; Xu et al., 2016) various scaling theories have been formulated to help identify the concentration regimes and also, for predicting the dependence of viscosity on concentration. Despite the numerous efforts were carried out on the intrinsic viscosity of polyelectrolyte solutions, it is still difficult to establish equations describing their viscometric behavior at different concentrations.

As per definition, the intrinsic viscosity ($[\eta]$) value of a given polymer in solution is the hydrodynamic volume of the polymer in the limit of infinite dilution and is normally obtained from intercept of the plots in the so-called Huggins plots (plot of the reduced viscosity, η_{red} *versus* the polymer concentration, c) (Huggins 1942). This method is well applied in the system containing uncharged polymers. Instead, in the case of polyelectrolytes, the electrostatic interactions between the charged groups leads to non-linear dependencies and a reliable determination of intrinsic viscosity becomes difficult. Recently, a purely phenomenological approach for the determination and interpretation of intrinsic viscosity, which avoided extrapolations and hence, complications of the above-mentioned methods, has been presented by Wolf (Wolf 2007; Eckelt et al., 2008). This contribution presents a novel and generally applicable alternative method for the determination of intrinsic viscosity from the initial slope of the dependence of the natural logarithm of the relative viscosity ($\ln \eta_{rel}$) on polymer concentration.

As mentioned, viscosity studies provide a reasonable evaluation of the bulk macroscopic solution behavior. Examination of viscosity data obtained by the Huggins and Wolf plots for solutions of polyelectrolytes, e.g., quaternized polysulfones, reveals the balance between the forces acting in polymeric complex systems over a concentration domain. In this context, Figure 1 gives indications on the electrolyte effect of quaternized PSFs by application of the Huggins equation (Huggins 1942) (Equation 1).

$$\eta_{sp} / c = [\eta]_{Huggins} + k_H [\eta]^2_{Huggins} c \tag{1}$$

where η_{sp} is the specific viscosity, k_H is the Huggins constant, and c is the concentration of polymer solution.

Thus, Figure 1a exemplifies the linear dependencies with positive slopes for neutral polymers, namely polysulfone (PSF) and chloromethylated polysulfones (CMPSF) in N,N-dimethylformamide (DMF) (Ioan et al., 2006a, 2006b; Filimon et al., 2007; 2010). Moreover, a decrease of η_{sp} / c with increasing polymer concentration - the specific behavior of the polyelectrolyte - is illustrated in Figure 1a for quaternized polysulfones with different ionic chlorine content (PSF-DMEtA1 and PSF-DMEtA2) in DMF/water (W) and methanol (MeOH)/W solvent mixtures (Filimon 2007). A clear polyelectrolyte behavior is also presented in Figure 1b for PSF-DMEA and PSF-DMOA in DMF and DMF/W mixtures (Filimon et al., 2010).

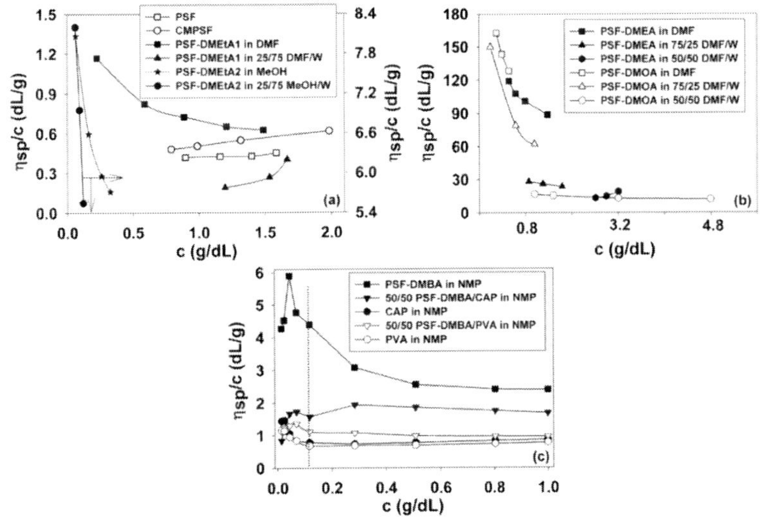

Figure 1. Huggins plots for uncharged polysulfones (PSF and CMPSF), and functionalized PSFs with different tertiary amine (DMEtA, DMEA, DMBA, DMOA - above-mentioned) in pure solvent and/or in solvent mixtures at 25°C.

Additionally, is shows the influence of solvent mixture composition on η_{sp}/c as depending on the neutral segment and on the charged groups (Figure 1a and b). Thus, the solvent/nonsolvent systems were selected on the one hand, as a function of the ionic chlorine content of quaternized polysulfones (e.g., PSF-DMEtA1 - Cl_i = 2.15% and PSF-DMEtA2 - Cl_i = 5.71%), and on the other hand, as a function of the alkyl chain length attached to the PSF backbone (e.g., PSF-DMEA - with two carbon atoms in the alkyl side chain, and PSF-DMOA -with eight carbon atoms in the alkyl side chain). Therefore, the composition domains of these mixtures were imposed by polymer solubility.

Additionally to the remarks above-mentioned, Figure 1c illustrates the influence of polymer mixtures composition on the η_{sp}/c. Thus, the increase of viscosity, which means an increase of coil dimension in solution, was generated by a higher content of charged groups from the alkyl radical of the quaternized polysulfones (e.g., DMBA) in the polymer mixtures (PSF-DMBA/PVA in N -methyl-2-pyrrolidone (NMP) and/or PSF-DMBA/CAP in NMP). This effect results from the higher electrostatic repulsions and cumulative steric hindrances, which caused a more expanded hydrodynamic volume and an increase of the long-range repulsive interactions.

In Figure 1c, secondly, is observed that in the region of low polymer concentrations (the values of relative viscosity lower that 1.2), a sharp increase followed by a rapid decrease of the reduced viscosity appears. According to literature (Eckelt et al., 2008), over an extremely dilute concentration regime, the viscosity abnormality can be observed, due to the adsorption phenomena produced on the viscometer wall surface. An important effect for polymers with polar groups (Hernandez-Fuentes et al., 1982; Filimon et al., 2014a; Bercea et al., 2017). For example, some recent studies obtained for polyvinyl alcohol solutions in water (Saeed et al., 2002; Yan et al., 2008) have evidenced that the solvent-solvent or polymer-polymer intramolecular interactions could determine adsorption of the polymer molecules on the viscometer capillary walls. Consequently, at very low polymer concentrations, the chains motion due to the flow of the solution or to Brownian motion can be explained by the interactions between chains and implicitly, the conformation changes of macromolecular chain could be attributed to sharp discontinuity observed. In this sense, for PSF-DMBA/CAP and PSF-DMBA/PVA mixtures, this maximum value is less evidenced than for pure PSF-DMBA. Moreover, the available volume is often higher than the volume occupied by the polymer molecules (Kulicke et al.,1989).

Generally, deviations from linearity reported in the extreme dilute and dilute concentration domains (Huggins equation) for charged polymer solutions are eliminated by means semi-emirical equations, like Rao (1993) or Wolf (Wolf et al., 2007; Eckelt et al., 2008). In particular, experimental data for quaternized polysulfones were fitted by applying the Rao (Equation 2) and/or Wolf (Equation 3) approximations for determination of the viscometer and hydrodynamic parameters (Ioan et al., 2006b; Filimon et al., 2007; 2010; 2014a).

$$\frac{1}{2\left(\eta_{rel}^{1/2} - 1\right)} = \frac{1}{[\eta]_{Rao}\, c} - \frac{(a - 1)}{2.5} \qquad (2)$$

$$\ln \eta_{rel} = \frac{c[\eta]_{Wolf} + Bc^2 [\eta]_{Wolf} [\eta]^{\bullet}}{1 + Bc[\eta]_{Wolf}} \qquad (3)$$

where the parameters involved are: $a = 1/\Phi_m$ and $\Phi_m = \frac{[\eta]}{2.5} c_m$ is the maximum volume fraction to which the particles can pack, B represents hydrodynamic interaction parameter, and $[\eta]^{\bullet}$ is the specific hydrodynamic volume.

The charged density and alkyl side chain size from the quaternized polysulfones (Figure 2a), and also solvent mixtures composition (Figure 2b) are parameters which influence of the values of intrinsic viscosity. Additionally, for multicomponent systems polyelectrolytes/polymer/ solvent (e.g., PSF-DMBA/PVA, PSF-DMBA/CAP – see Figure 2c), the dependence of intrinsic viscosity on polymer mixtures composition is complex. This aspect it is reflected in the conformational changes of each polymer within mixing, and also in the thermodynamic or hydrodynamic interactions among polymers. Accordingly, the values of $\lfloor\eta\rfloor$ are affected by the charge density of the alkyl radical of the quaternized polysulfone, hydrophilic characteristics of the neutral polymers (PVA or CAP), composition of the polymer mixtures, and also by the used solvent.

On the other hand, the dependence of intrinsic viscosity on temperature is interpreted in terms of a conformational change of the polymer chain. The prevailing factor is the polarity of solvent in solvation power. Therefore, for a given composition of the solvent/nonsolvent mixtures (e.g., DMF/MeOH, DMF/water, or MeOH/water), one of the components is preferentially adsorbed to the quaternized PSF molecules, in the direction of a thermodynamically most effective mixture.

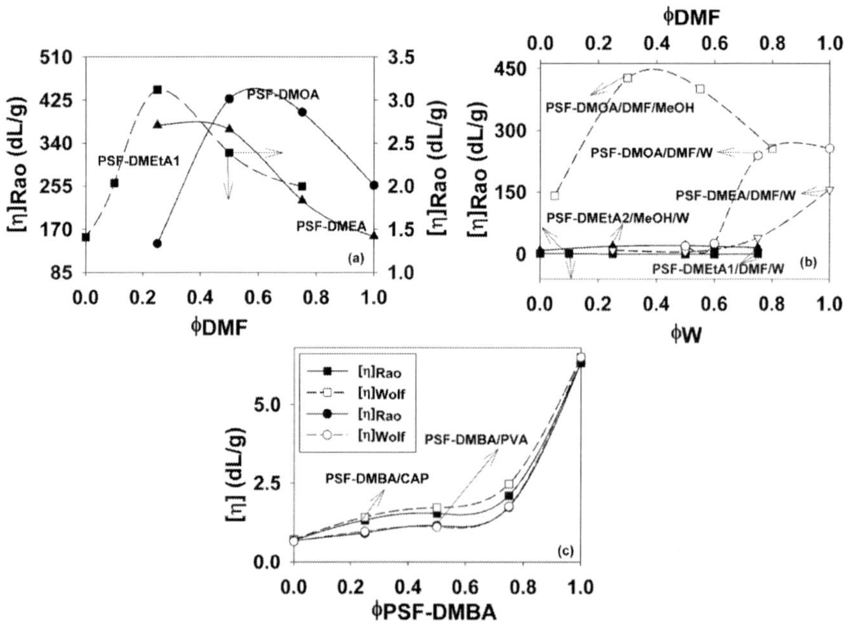

Figure 2. Dependence of intrinsic viscosity (evaluated by Rao or Wolf method) on the composition of solvent/nonsolvent mixtures (a, b) and also, on the volume fraction of polyion in case of polymer mixtures (c), at 25°C.

Interactions in Ionic Polymer Systems: A Predictive Model for the Conformation and Its Concentration Dependences

Strength of the interactions between components of a polymeric system represents the key factor in polymer materials characterization. The experimental investigations and theoretical approximations was used for underlying the establishment of some properties prior to the synthesis process, and for elucidating some aspects concerning the behavior of macromolecular compounds in different fields (Vebber et al., 2014; Milczewska et al., 2014). In this context, many existent theoretical models have focused on finding the structure–property relationships in order to predetermine some structures with specific applications. Consequently, the control of polymeric structure is among the most important targets of modern macromolecular science. In particular, tailoring the position and strength of interaction forces within macromolecules by synthetic methods and therefore, structural control on the final polymeric materials have become possible.

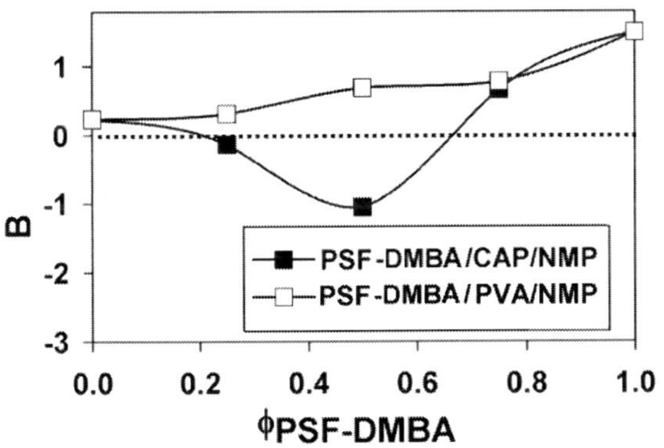

Figure 3. Dependence of hydrodynamic interaction parameter, B, on the volume fraction of PSF-DMBA.

As known, charged macromolecules introduce new lenght scales, such as the Debye screening length and Bjerrum lenght. Moreover, the long-range Coulomb interactions lead to a variety of effects qualitatively different from those of neutral polymers with respect to, for example, their viscoscometric behavior, thermodynamic, equilibrium, and transport properties. This thermodynamic aspect becomes particularly important; the favorable or unfavorable interactions govern the probability of contact formation between the polymer segments and solvent molecules and therefore, is greatly influenced the conformation of macromolecular chains in the solution.

According to these considerations, based on viscometric data, is emphasizes the importance the competition among different thermodynamic interactions on conformational properties, reffering in particular to the ionic polysulfones in dilute solution and polymer mixtures (see Figure 3).

Generally, the literature (Kulicke et al., 1989) indicates that, the hydrodynamic interaction parameter, B, quantifies the interactions between the polymer segments belonging to different macromolecules. In addition, literature underlines that, in the case of polyelectrolyte solutions, the B parameter involves two parts, one comprising all effects of Coulomb interactions ($B_{Coulomb}$ - represents the positive contribution of Coulomb interactions to the hydrodynamic interaction parameter) and another part containing interactions that are present in neutral polymers ($B_{non\ Coulomb}$ - is a negative contribution of non-Coulomb interactions to the hydrodynamic interaction parameter) (Eckelt et al., 2008) (Equation 4).

$$B = B_{Coulomb} + B_{non\text{-}Coulomb} \tag{4}$$

Modification of the polymer coil dimension, and implicitly, the chain conformation changes can be attributed the excluded volume effects as a result of the repulsive intermolecular interactions between ionic PSFs and CAP or PVA in NMP. Moreover, addition of CAP or PVA to the solution of PSF-DMBA favors intramolecular and polymer-solvent interactions, while the CAP or PVA chains dissolved in NMP exhibit the ability to develop interactions of the hydrogen bonds type (Yang 2000; Filimon et al., 2014a), contributing to the specific molecular rearrangement of the system.

It is noteworthy that, the chain conformation of polyelectrolytes is mainly influenced by their charge density and counterion dissociation. This changing trend is visualised in Figure 3. In this context, the presence of electrostatic interactions, reflected in the significant contribution of Coulomb interaction, $B_{Coulomb}$, leads to high and positive B values. On the other hand, the addition of uncharged polymers (e.g., PVA or CAP) to ionic PSFs solutions in different mixing ratio determines a counterbalance of Coulomb interactions through non-Coulomb interactions up to B becomes zero. Over the entire range of composition the positive values of B parameter were obtained only for the PSF-DMBA/PVA system. Instead, for PSF-DMBA/CAP system, at a higher amount of CAP (over 0.5 volume fraction) the negative values of B parameter were observed, which suggests the manifestation of an aggregation tendency, as results of the long-range interactions between polymer chains. The origin of these long-range interactions is not due to the extended or stretched conformation of the individual polyelectrolyte chains, but it is rather due to the Coulomb interactions or aggregates formation that contributes to this electrostatic behavior (Kulicke 1989; Filimon et al., 2014a).

According to Wolf (2007; Eckelt et al., 2008), the high degree of coil shrinkage caused by the increase of polymer concentration becomes directly visible in terms of the generalized specific hydrodynamic volume $\lim_{\substack{c \to 0 \\ \dot{\gamma} \to 0}} \{\eta\} = [\eta]$.

In this context, the specific hydrodynamic volume changes are caused by the progressive screening of the electrical charges resulting from the additional molecules. The observed behavior can be explained it taking into account that the global conformation of polymer mixtures chains is influenced by the charge density of the polyion, concentration, and polyelectrolyte/neutral polymer mixtures composition.

Consequently, the behavior of polyelectrolytes in solution or in mixture with other polymers can provide important information about the electrostatic interactions useful for practical activities. Thus, in terms of practical importance, these systems based on ionic PSFs, may be applied in the production of smart materials with excellent properties for obtaining a particular morphology, biocompatibility, etc. for a large variety of structural arrangements in biomedical applications.

SOLUTION DYNAMICS STUDY OF IONIC POLYMERS: TAILORING OF PERFORMANCE PROPERTIES

The design of novel polymeric structures having specific macroscopic properties such as texture, appearance, chemical and microbiological stability, and pharmacological of physiological effectiveness continues to be a major challenge for scientific fields. Structural design approaches based on knowledge concerning the conformational properties of polymers in solution, as well as fundamental knowledge of the interaction between molecules and also, the interaction between the structural entities subsequently created leads to the formation of structure with specific macroscopic performances (Higuchi et al., 1992; Nady et al., 2011). In this context, the molecular structure and conformation of polymers are important parameters in the study of solution dynamics of polymers/polyelectrolytes, as allowing a fine tuning and control of the flow properties of fluids, which has a great importance in various areas, e.g., biomedical, pharmaceutical, cosmetic fields (Filimon et al., 2016). Additionally, the dependence between the structure of individual molecules in solution and molecular parameters is essential to understand microstructural changes in the polymer materials with important role in the obtaining processes of membranes with controlled dimensions for different applications (Higuchi et al., 1992; Nady et al., 2011).

Consequently, dynamic study is one of the most effective tools to investigate the flow and viscoelastic behavior of the polymer solution processing, reflecting the cumulative effects generated by molecular structure of polymers, type of interactions and/or composition of solvent mixtures/polymer mixtures. Therefore, the solution dynamics study of polyelectrolyte systems is necessary to gain insight into many biological and industrial processes. From theses considerations, the ionic polysulfones have received widespread attention for their promising roles as ionic exchange membranes, antibacterial coatings, and dialysis membranes.

Conformational restructurations generated by electrostatic interactions and hydrogen bonding occurring in polyelectrolytes/solvent/nonsolvent systems have been evidenced by rheological properties (Filimon et al., 2013). To describe the dynamic viscosity of polymers, literature (Li et al., 2008) indicated the existence of three shear flow regimes, such as: Region I - where apperas a shear thinning behavior at low shear rates, Region II - characterised by a Newtonian plateau at intermediate shear rates, and another shear thinning regime at high shear rates - Region III. In this respect, Figure 4 exemplifies the flow behavior of some functionalized PSFs with different pendant groups in the pure solvent and/or solvent/nonsolvent mixtures observed from the dynamic viscosity-shear rate dependence.

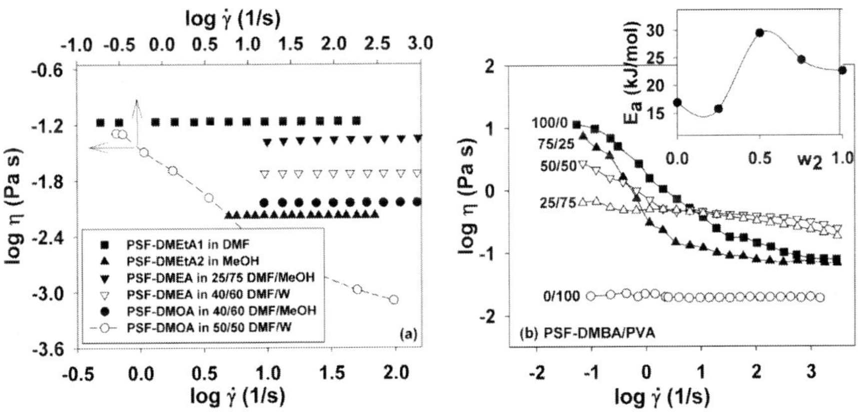

Figure 4. Double logarithmic plot of dynamic viscosity *versus* shear rate for: (a) quaternized PSFs with different radical alkyl (DMEtA, DMEA, DMOA) in DMF MeOH, DMF/MeOH, and DMF/water mixtures; (b) for PSF-DMBA/PVA blend at various mixing ratios, at 25°C. Inserted plot illustrates variation of activation energy as a function of PVA weight fraction (w_2) for PSF-DMBA/PVA blend.

Thus, a Newtonian behavior were obtained for PSF-DMEtA1 and PSF-DMEtA2 in DMF, MeOH, and DMF/MeOH solvent mixtures (Figure 4a). Instead, the lower flexibility of PSF-DMOA, compared with that of PSF-DMEA, and the different solvent mixtures (e.g., DMF/MeOH and DMF/water), influence the dynamic viscosity (see Figure 4a). In this context, all studied solutions present a constant viscosity region on the studied shear rates domain (a Newtonian behavior), except for the sample PSF-DMOA in DMF/water. In this last case, the shear thinning is observed at low shear rates according to the studies of Witten and Cohen (Witten and Cohen 1985). At the same time, it is observed that, as shear rate increases, the slope of curves decreases and a Newtonian plateau appears. This plateau shrinks and moves toward higher shear rates as water composition increases.

Recent studies (Lu et al., 2013; Filimon et al., 2016) showed that the investigation of the rheological properties of pure components is essential for a better understanding of the viscoelastic properties developed by quaternized PSFs/synthetic polymer blends (i.e., PSF-DMEA, PSF-DMBA or PSF-DMOA blends with PS, PVA, CAP, and/or P4VP). These combined studies contribute to the knowledge on the nature of the interactions manifested in multicomponent system and also, allow modulation of some the functional properties by creating new materials. Thus, Figure 4b reveals the complex flow behavior of PSF-DMBA blended PVA in NMP at different ratios. Examination of flow curves (Figure 4b) indicates that, according to literature (Li et al., 2008), the dynamic viscosity - shear rate dependence appears only in Regions I and II. Thus, the ionic polysulfones (PSF-DMBA) and blends with a high content of PSF-DMBA (over 0.5 weight fraction) exhibit a shear thinning behavior at low shear rate (Region I), followed by a Newtonian plateau, as shear rate increases (Region II). In other words, the different flow behavior can be explained by considering the synergism between the structural properties of the blend polymers (the charge density and size of polycation), flexibility of molecular backbone which, in turn, is expected to depend on their chain packing efficiency and specific interactions, their mixing rations, and the solvent used. One can conclude that these blends possess properties of pseudoplastic materials and the variation of PVA composition in studied system generates the specific molecular rearrangement of the system, suggesting that it acts as a plasticizer.

The different radical alkyl from quaternized PSFs influence the solubility process in a certain solvent or different solvent mixtures, generating molecular restructurations in solution, under the influence of hydrogen bonding, electrostatic interactions, and association phenomena. Thus, one can mention that the rheological properties are influenced by the hydrophobic/hydrophilic characteristics of the quaternized PSFs, which decrease in the following order: PSF-DMOA > PSF-DMBA > PSF-DMEA > PSF-EtA. Moreover, for a multicomponent system, quaternized PSFs/neutral polymer/solvent, the effect of the combined action of different thermodynamic interactions from the system, as well as the conformational changes of each type of polymer from the blends modifies the flow activation energy, E_a (Arrhenius equation) (de Vasconcelos et al., 2002). The decrease of the energetic flow barrier (E_a) is generated by the polyelectrolyte effect and the increase of energetic flow barrier shows the influence of the different flexibility degrees of the polymeric chains, and also, the competition between the intra- and intermolecular interactions. This behavior reflects the significant contribution of the electrostatic interactions and also, suggests the manifestation of an aggregation tendency as results of the long-range interactions between the polymer chains.

Physical chain entanglements behave similar to chemical cross-links, although the chains can also slide one over another, affecting the viscoelastic behavior. This aspect which completes the studies developed in shear regime and which provides insight into dynamics and architecture of chain (Wood-Adams et al., 2000) is reflected in the storage, G', and loss, G'', moduli. In this context, some studies (Filimon et al., 2013; 2016) have discussed the the effect of polymers structural characteristics and blends composition on the mobility of chains in the shear field, and also the response of shear moduli on oscillation frequency range, f. Figure 5 exemplifies the variation of storage and loss moduli *versus* frequency for quaternized PSFs (e.g., PSF-DMEtA1 and PSF-DMEA) at different compositions of mixed solvents. Also, the rheological response for PSF-DMBA/PVA blends over the oscillation frequency range respects the power law and showed a typical behavior characteristic to viscoelastic Maxwellian fluids (Gupta and Yasee 1997) (Figure 6).

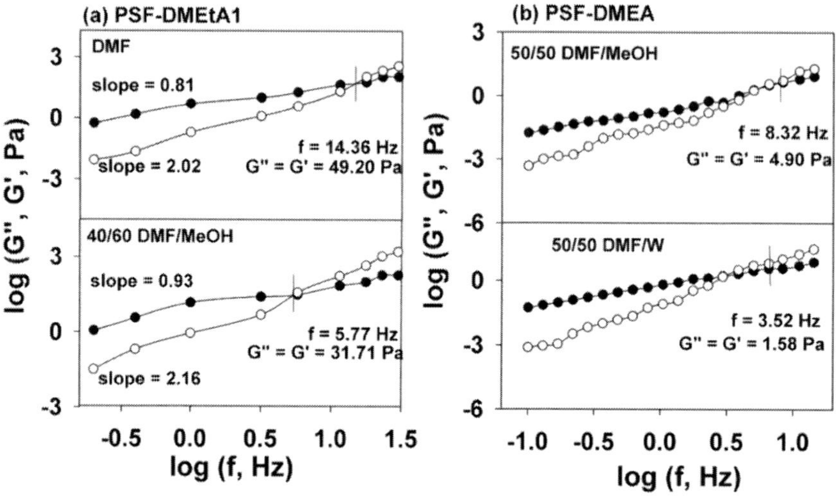

Figure 5. Double-logarithmic plots of shear moduli *versus* oscillatory frequency for PSF-DMEtA1 (a) and PSF-DMEA (b) at different composition of solvent/nonsolvent mixtures, and 25°C.

Moreover, as shown in Figure 5 and 6, the conformational changes of the polymer and the cumulative effects of electrostatic interactions, hydrogen bonding or aggregation phenomena are viewed in the values of the crossover point, $f_{G'=G''}$. All these results - correlated with those obtained from the viscosity - reflect the specific molecular rearrangements, some associations, and structural phenomena in the multicomponent polyelectrolyte systems.

Consequently, knowledge of the solution properties of ionic PSFs becomes important for establishing the correlation with specific morphological aspects, in the order to determine the proper films with modeled morphology which represent the basis to future applications, e.g., as membrane designed for biomedical field.

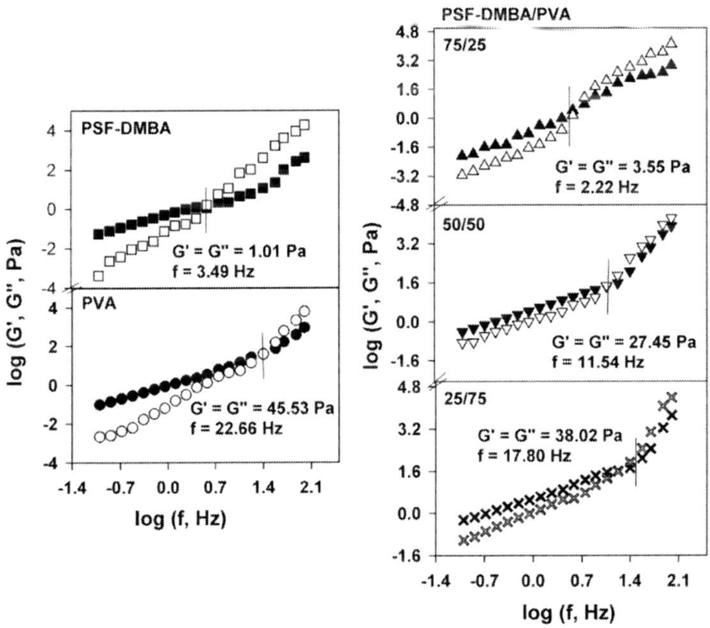

Figure 6. Double-logarithmic plots of the storage, G' (open symbols), and loss, G" (solid symbols) moduli *versus* oscillatory frequency, f, for PSF-DMBA/PVA blend at various mixing ratios (wt./wt.) and 25°C.

REFERENCES

Avram, E. (2001). Polymers with pendent functional groups. VI. A comparative study on the chloromethylation of linear polystyrene and polysulfone with paraformaldehyde/Me_3SiCl. *Polymer-Plastics Technology and Engineering*, 40: 275–281. http://dx.doi.org/10.1081/PPT-100000248.

Avram, E. and Filimon, A. (2015). Aromatic linear polysulfones with pendant functional groups. In *Functionalized Polysulfones: Synthesis,*

Characterization, and Applications, edited by Ioan, S., 3-35. Taylor and Francis, CRC Press.

Barikani, M. and Shahram, M.-A. (2000). Synthesis, characterization and thermal properties of novel arylene sulfone ether polyimides and polyamides. *Polymer Science Part A: Polymer Chemistry*, 38: 1487-1492. doi:10.1002/(SICI)1099-0518(20000501)38: 9<1487::AID-POL A11>3.0.CO;2-U.

Bercea, M., Morariu, S. and Wolf, B. A. (2017). Consequences of linking charged and uncharged monomers to binary copolymers studied in dilute solution. Part I: Viscometric behavior of the homopolymers, the effects of charging, and uncommon salt effects. *European Polymer Journal*, 88: 412–421. https://doi.org/10.1016/j.eurpolymj.2016.12.008.

Bettinger, C. J. (2011). biodegradable elastomers for tissue engineering and cell–biomaterial interactions. *Macromolecular Bioscience*, 11: 467-482. doi:10.1002/mabi.201000397.

Chandima, J. K. and Jagath, P. K. (2007). Names, acronyms, classes, and structures of some important polymers. In *Physical Properties of Polymers Handbook*, *Second Edition*, editd by Mark, E. J., 25-40. Cincinnati: Springer.

de Gennes, P.-G., Pincus, P., Velasco, R. M. and Brochard, F. (1976). Remarks on polyelectrolyte conformation. *Le Journal de Physique*, 37: 1461-1473. doi:10.1051/jphys:0197600370120146100.

de Vasconcelos, C. L., Martins, R. R., Ferreira, M. O., Pereira, M. R. and Fonseca, J. L. C. (2002). Rheology of polyurethane solutions with different solvents. *Polymer International*, 51: 69–74. doi:10.1002/pi.800.

Dizman, C., Tasdelenc, M. A. and Yagcia, Y. (2014). Recent advances in the preparation of functionalized polysulfones. *Polymer International*, 62: 991–1007. doi:10.1002/pi.4525.

Dobrynin, A. V., Colby, R. H. and Rubinstein, M. (1995). Scaling theory of polyelectrolyte solutions. *Macromolecules*, 28: 1859–1871. doi:10.1021/ma00110a021.

Dobrynin, A. V. and Rubinstein, M. (2005). Theory of polyelectrolytes in solutions and at surfaces. *Progress in Polymer Science*, 30: 1049–1118. doi:10.1016/j.progpolymsci.2005.07.006.

Du, R.. and Zhao, J. (2004a). Propertics of poly (N,N-dimethylaminoethyl methacrylate)/polysulfone positively charged composite nanofiltration membrane. *Journal of Membrane Science*, 239: 183–188. https://doi.org/10.1016/j.memsci.2004.03.029.

Du, R. and Zhao, J. (2004b). Positively charged composite nanofiltration membrane prepared by poly(N,N-dimethylaminoethyl methacrylate)/polysulfone. *Journal of Applied Polymer Science*, 91: 2721–2728. doi:10.1002/app.13477.

Eckelt, J., Knopf, A. and Wolf, B. A. (2008). Polyelectrolytes: Intrinsic viscosities in the absence and in the presence of salt. *Macromolecules*, 41: 912-918. doi:10.1021/ ma 702054f.

Filimon, A., Avram, E. and Ioan, S. (2007). Influence of mixed solvents and temperature on the solution properties of quaternized polysulfones. *Journal of Macromolecular Science, Part B: Physics*, 46: 503–520. doi:10.1080/00222340701257752.

Filimon, A., Avram, E., Dunca, S., Stoica, I. and Ioan, S. (2009). Surface properties and antibacterial activity of quaternized polysulfones. *Journal of Applied Polymer Science*, 112: 1808–1816. doi:10.1002/app.29591.

Filimon, A., Albu, R. M., Avram, E. and Ioan, S. (2010). Effect of alkyl side chain on the conformational properties of polysulfones with quaternary groups. *Journal of Macromolecular Science, Part B: Physics*, 49: 207–217. doi:10.1080/ 002223 40903346494.

Filimon, A., Avram, E. and Ioan, S. (2013). Structure-rheology relationship in complex quaternized polysulfones/solvent/nonsolvent systems. *Polymer Bulletin*, 70: 1835–1851. doi:10.1007/s00289-012-0874-z.

Filimon, A., Dobos, A. M., Avram, E. and Ioan, S. (2014a). Ionic polymers based on quaternized polysulfones: hydrodynamic properties of polymer mixtures in solution. *Pure and Applied Chemistry*, 86: 1871-1882. https://doi.org/10.1515/pac-2014-0603.

Filimon, A., Avram, E. and Stoica, I. (2014b). Rheological and morphological characteristics of multicomponent polysulfone/ poly(vinyl alcohol) systems. *Polymer International*, 63: 1856–1868. doi:10.1002/pi.4716.

Filimon, A., Avram, E. and Dunca, S. (2015). Surface and interface properties of functionalized polysulfones: cell-material interaction and antimicrobial activity. *Polymer Engineering and Science*, 55: 2184-2194. doi:10.1002/pen.24103.

Filimon, A., Albu, R. M., Avram, E. and Stoica, I. (2016). Blends based on ionic polysulfones with improved conformational and microstructural characteristics: Perspectives for biomedical applications. *Composites Part B: Engineering*, 93: 1-11. https://doi.org/10.1016/j.compositesb.2016.02.062.

Gupta, K. and Yaseen, M. (1997). Viscosity-temperature relationship of dilute solution of poly(vinyl chloride) in cyclohexanone and in its blends with xylene. *Journal of Applied Polymer*, 65: 2749–2760. doi: 10.1002/(SICI)1097-4628(19970926)65,13<2749::AID-APP18>3.0.CO;2-S.

Higuchi, A., Koga, H. and Nakagawa, T. (1992). Surface-modified polysulfone hollow fibers. iv. chloromethylated fibers and their derivatives. *Journal of Applied Polymer Science*, 46: 449-457. doi:10.1002/app.1992.070460310.

Hernandez-Fuentes, I., Prolongo, M. G., Masegosa, R. M. and Horta, A. (1982). Limits of the dilute regime for the solution viscosity of pmma in good and in poor solvents. *European Polymer Journal*, 18: 29-35. https://doi.org/10.1016/0014-3057(82)90128-8.

Huggins, M. L. (1942). The viscosity of dilute solutions of long-chain molecules. IV. Dependence on concentration. *The Journal of the American Chemical Society*, 64: 2716–2718. doi:10.1021/ja01263a056.

Ioan, S., Filimon, A. and Avram, E. (2006a). Influence of the degree of substitution on the solution properties of chloromethylated polysulfone. *Journal of Applied Polymer Science*, 101: 524–531. doi:10.1002/app.23340.

Ioan, S., Filimon, A. and Avram, E. (2006b). Conformational and visometric behavior of quaternized polysulfone in dilute solution. *Polymer Engineering and Science*, 46: 827–836. doi:10.1002/pen.20526.

Karlsson, L. E. and Jannasch, P. (2004). Polysulfone ionomers for proton-conducting fuel cell membranes: sulfoalkylated polysulfones. *Journal of Membrane Science*, 230: 61–70. https:// doi.org/10.1016/j.memsci.2003.10.033.

Kotlyarevskaya, O. O., Navrotskii, V. A., Orlyanskii, M. V., Navrotskii, A. V. and Novakov, I. A. (2005). Hydrodynamic behavior of N,N-dimethylaminoethyl methacrylate polylectrolytes in mixed solvents. *Polymer Science. Series A*, *Chemistry*, *Physics*, 47: 313–318.

Kulicke, W. M., Kotter, M. and Grager, H. (1989). Drag reduction phenomenon with special emphasis on homogeneous solutions." *Advances in Polymer Science*, 89: 1-68.

Kuroiwa, T., Miyagishi, T., Ito, A., Matsuguchi, M., Sadaoka, Y. and Sakai, Y. (1995). A thin-film polysulfone-based capacitive-type relative-humidity sensor. *Sensors and Actuators B: Chemical*, 25: 692–695. https:// doi.org/ 10.1016/0925-4005(95)85153-4.

Li, H., Shen, X.-Y., Gong, G.-L. and Wang, D. (2008). Compatibility studies with blends based on hydroxypropylcellulose and polyacrylonitrile. *Carbohydrate Polymers*, 73: 191-200. https:// doi.org/10.1016/j.carbpol.2007.11.022.

Liu, M., Xiao, C. and Hu, X. (2011). Effects of non-solvent additives on pvdf solution viscosity and membrane performance. *Iranian Polymer Journal*, 20: 979-988.

Lu, F., Song, J., Cheng, B.-W., Ji, X.-J. and Wang, L.-J. (2013). Viscoelasticity and rheology in the regimes from dilute to concentrated in cellulose 1-ethyl-3-methylimidazoliumacetate solutions. *Cellulose*, 20: 1343–1352. doi:10.1007/s10570-013-9885-7.

Manea, C. and Mulder, M. (2002). Characterization of polymer blends of polyethersulfone/sulfonated polysulfone and polyethersulfone/sulfonated polyetheretherketone for direct methanol fuel cell applications. *Journal of Membrane Science*, 206: 443–453. https://doi.org/10.1016/S0376-7388(01)00787-6.

Milczewska, K., Voelkel, A. A. and Piedzia, K. (2014). Interactions in PEG/aerosil and PLA/aerosil composites described by IGC-determined Flory-Huggins χ_{23} parameter. *Journal of Polymer Research*, 21: 394–400. doi:10.1007/s10965-014-0394-7.

Muthukumar, M. (1997). Dynamics of polyelectrolyte solutions. *The Journal of Chemical Physics*, 107: 2619. doi: http://dx.doi.org/10.1063/1.474573.

Nady, N., Franssen, M. C. R., Zuilhof, H., Eldin, M. S. M., Boom, R. and Schroën, K. (2011). Modification methods for poly(arylsulfone) membranes: A mini-review focusing on surface modification. *Desalination*, 275: 1–9. https://doi.org/ 10. 1016/j.desal.2011.03.010.

Odijk, T. (1979). Possible scaling relations for semidilute polyelectrolyte solutions. *Macromolecules*, 12: 688-693. doi:10.1021/ ma60070a028.

Rao, M. V. S. (1993). Viscosity of dilute to moderately concentrated polymer solutions. *Polymer*, 34: 592-596. https://doi.org/10.1016/0032-3861(93)90555-O.

Saeed, R., Uddin, F. and Fazal, A. (2002). Effect of electrolyte concentration on viscous flow of polymer solutions. *Journal of Chemical Engineering Data*, 47: 1359–1362. doi: 10.1021/ je020022w.

Sanchez, S., Pumera, M., Cabruja, E. and Fàbregas, E. (2007). Carbon nano tube/polysulfone composite screen printed electrochemical enzyme biosensor. *Analyst*, 132: 142–147. doi:10.1039 /B609137G.

Sivaraman, K. M., Kellenberger, C., Pané, S., Ergeneman, O., Lühmann, T., Luechinger, N. A., Hall, H., Stark, W. J. and Nelson, B. J. (2012). Porous polysulfone coatings for enhanced drug delivery. *Biomedical Microdevices*, 14: 603–612. doi: 10.1007/s10544-012-9639-6.

Vebber, G. C., Pranke, P. and Pereira, C. N. (2014). Calculating Hansen solubility parameters of polymers with genetic algorithms. *Journal of Applied Polymer Science*, 131: 39696–39707. doi:10.1002/app.39696.

von Kraemer, S., Puchner, M., Jannasch, P., Lundblad, A. and Lindbergh, G. (2006). Gas diffusion electrodes and membrane electrode assemblies based on a sulfonated polysulfone for high-temperature PEMFC. *Journal of the Electrochemical Society*, 153: A2077–2084.

Wood-Adams, P. M., Dealy, J. M., deGroot, W. A. and Redwine, D. O. (2000). Effect of molecular structure on the linear viscoelastic behavior of polyethylene. *Macromolecules*, 33: 7489-7499. doi:10.1021/ ma991 533z.

Wolf, B. A. (2007). Polyelectrolytes revisited: Reliable determination of intrinsic viscosities. *Macromolecular Rapid Communications*, 28: 164-170. doi:10.1002/marc.200600650.

Witten, T. A. and Cohen, M. H. (1985). Crosslinking in shear-thickening ionomers. *Macromolecules*, 18: 1915–1918. doi:10.1021/ma 00152a019.

Xu, G., Luo, S., Yang, Q., Yang, J. and Zhao, J. (2016). Single chains of strong polyelectrolytes in aqueous solutions at extreme dilution: Conformation and counterion distribution. *The Journal of Chemical Physics*, 145: 144903. doi: http://dx.doi.org/10.1063/1.4964649.

Yan, Y.-F., Li, H.-Z., Yang, H.-Y., Qian, J.-S., Zhu, P.-P. and He, P.-S. (2008). Effect of short chain alcohols upon viscosity of ttab solution. *Chinese Journal of Chemical Physics*, 21: 169. doi: http://dx.doi. org/10.1088/1674-0068/21/02/169-173.

Yang, H., Zhu, P., Ren, F., Wang, Y. and Zhang, T. (2000). Viscometric investigations on the intermolecular interactions between poly(methyl methacrylate) and poly(vinyl acetate) in various solvents. *European Polymer Journal*, 36: 21-26. https://doi.org/10.1016/S0014-3057 (99)00049-X.

In: Solution Chemistry
Editor: Yongliang Xiong
ISBN: 978-1-53613-101-7

Chapter 6

DETERMINATION OF TRACE ELEMENTS IN CRUDE OILS AND FUEL OILS: A COMPREHENSIVE REVIEW AND NEW DATA

Weihang Yang**, *Yongjun Gao* and *John F. Casey
Department of Earth and Atmospheric Sciences,
University of Houston, Houston, TX, US

ABSTRACT

Multiple trace elements in crude oils are of great importance in helping understanding oil formation, oil-oil correlations, oil-source rock correlations, oil migration, maturation of organic matters, and the depositional environment of source rocks, deciding if specific trace metal elements should be removed prior to refinery, and understanding if high abundances of toxic trace metal elements are doing harm to the local environment. Crude oil is a very complex mixture of both organic and inorganic matters. It requires pretreatment (sample preparation) prior to introdued into analytical instruments. Direct introduction and mineralization are the most widely used sample preparation techniques.

* Corresponding Author Email: wyang6@uh.edu.

Direct introduction includes direct introduction of crude oil after dilution with organic solvent and direct emulsion/micro-emulsion introduction. Mineralization includes strong acid digestion and ashing, or wet or dry combustion. For microwave assisted strong acid digestion, SRC (single reaction chamber) systems are more efficient in determining trace elements than rotor-based systems. FAAS (flame atomic absorption spectrometry), ET-AAS (electrothermal atomic absorption spectrometry) or GF-AAS (graphite furnace atomic absorption spectrometry), CVG-AAS (chemical vapor generation atomic absorption spectrometry), ICP-OES (inductively coupled plasma optical emission spectrometry), and ICP-MS (inductively coupled plasma mass spectrometry) are the most commonly used techniques to determine trace elements in crude oil and fuel oil samples. Every technique has its own pros and cons. A more advanced technique, QQQ-ICP-MS (triple quadrupole inductively coupled plasma mass spectrometry), was pioneered a couple of years ago, which helped to solve problems of the typical single quadrupole ICP-MS (Q-ICP-MS), such as polyatomic interferences, isobaric interferences, and non-mono-charged interferences.

Keywords: multiple trace elements, crude oils, fuel oils, sample preparation, determination, direct introduction, mineralization, rotor-based microwave, SRC microwave, FAAS, ET-AAS (GF-AAS), CVG-AAS, ICP-OES, ICP-MS, QQQ-ICP-MS

INTRODUCTION

Crude oil is a complex mixture of hydrocarbons of various molecular weights and other organic compounds, formed from plants and animals that lived millions of years ago [1, 2]. It is yellow to black liquid found in geological formations beneath the earth surface, commonly separated into components to provide various types of fuels using a technique called fractional distillation [1]. Petroleum covers both naturally occurring unprocessed crude oil and petroleum products that are made up of refined crude oil. It is a fossil fuel, formed when large quantities of dead organisms, typically zooplankton and algae, are buried underneath sedimentary rock and subjected to both intense heat and pressure [1].

Carbon and hydrogen are the most abundant elements in crude oils, ranging from 82-87% and 12-15%, respectively [3]. Sulfur is the third most abundant element in crude oil, ranging from 0.05% to more than 5%. Oxygen and nitrogen are the fourth and fifth most abundant elements in crude oil, respectively. Generally, oxygen contents are less than 2%, and nitrogen contents are less than 0.1% [4]. The other elements (e.g., vanadium, nickel, calcium, potassium, iron, copper, zinc, boron, arsenic, selenium, silicon, phosphorus, etc.) are in trace level (i.e., 0.01-0.1%) or ultra-trace level (i.e., < 0.01%).

Although crude oils derived from different sources may be different from each other [5], they are based on the same three basic chemical series: paraffins, naphthenes, and aromatics [3]. Paraffins and naphthenes are saturated, and the former ones have chain-like structures, while the latter ones are closed ring series. Paraffins comprise the most common hydrocarbons in crude oil, and naphthenes are important in all liquid refinery products [3]. Different from paraffins and naphthenes, aromatics have unsaturated closed ring structures. They constitute a small fraction of all crude oils. Being the most common aromatic series, benzene exists in most of the crudes [3]. Fractionation and isolation schemes of crude oil are based on crude oil type [6]. One of the most commonly used fractionation schemes is SARA, which is composed of the capitals of the names of fractions produced, namely saturates (S), aromatics (A), resins (R), and asphaltenes (A). Crude oil separation and identification including SARA method have been reviewed by Mansfield et al., (1999) [7]. Crude oil separation and characterization have been carried out by a number of researchers [8, 9, 10, 11, 12, 13, 14, 15]. Asphaltenes comprise polar fractions of crude oils, including polyaromatics and heteroaromatics, and most trace metal elements are present in asphaltenes [16].

TRACE METAL ELEMENTS IN CRUDE OILS

The porphyrins and non-porphyrins of Ni (II) and VO (II) are the most common metals existing in crude oils. Other metal ions include lithium, beryllium, sodium, magnesium, aluminum, potassium, calcium, scandium, titanium, chromium, manganese, iron, cobalt, copper, zinc, gallium, germanium, rubidium, strontium, yttrium, zirconium, niobium, molybdenum, palladium, silver, cadmium, tin, antimony, caesium, barium, lanthanum, cerium, praseodymium, neodymium, samarium, europium, gadolinium, terbium, dysprosium, holmium, erbium, thulium, ytterbium, lutetium, hafnium, tantalum, tungsten, platinum, gold, mercury, thallium, lead, bismuth, thorium, and uranium [17, 18, 19, 20, 21, 22, 23, 24, 25, 26, 27, 28, 29, 30, 31, 32, 33, 34, 35]. Widely distributed and highly concentrated chlorophylls are the most common precursors of porphyrins [36]. Etioporphyrins (etio) and dexophylloerithroetioporphyrins (DPEP) and their homologues are more abundant among the porphyrins found in crude oils [37, 38]. Porphyrins can be classified into two groups. One is derived from chlorophylls of algae and phytoplanktons, mostly present in marine depositional environments. The other one is transformed from plant chlorophylls, which indicate a terrestrial origin [36]. Conversion of chlorophylls to porphyrins involves a series of reactions in the water column and during burial in sedimentation [39, 40, 41, 42, 43]. Filby (1994) [44] directly concluded that metalloporphyrins are formed during early diagenesis and catagenesis of the source rocks. Although the isolation of metalloporphyrins is not that easy due to the complexity of their mixtures, the separation and identification have been reported by enhanced spectroscopic and chromatographic techniques [45]. The V and Ni porphyrins were identified much earlier [36], and the organic forms of other metal elements were obtained later by gas chromatography (GC) or high performance liquid chromatography (HPLC) coupled to inductively coupled plasma – mass spectrometry (ICP-MS) or atomic absorption spectrometry (AAS). For example, the porphyrins of Ti, Zn, Cr, and Co were distinguished by HPLC-ICP-MS in oil shales [45].

The ratios of V/(Ni+V) or V/Ni were found to be constant in crude oils of common rock source, and up to the rock's geologic age [46]. The ratios can be used for characterization of source rocks [47, 48]. The compounds of VO DPEP to etio were proposed to be biomarkers due to their maturity dependence revealed by their thermal evolution investigations [35, 49, 50, 51].

The structures of metalloporphyrins are characterized by tetrapyrrolic complexes that are similar to heme and chlorophyll [36]. Nonporphyrins are polar compounds, and mostly present as cations of organic acids, but their chemical nature requires further studies [3]. Sulfur, nitrogen, and oxygen can be constituents of various nonporphyrins [52]. Trace elements can also be present in associated formation waters or mineral substances. Large quantities of trace metal elements may exist in humic substances, which have large capacities for metal complexation [53]. Kerogen may be derived from such humic substances with trace metal elements [3].

Determination of Trace Elements in Crude Oils

Trace element mass fractions and ratios in crude oils are quite helpful in understanding oil formation, oil-oil correlations, oil-source rock correlations, oil migration, maturation of organic matters, and the depositional environment of source rocks [17, 20, 54, 55, 56, 57, 58, 59]. Additionally, some trace metal elements act as catalyst poisons during the catalytic cracking process of crude oils in the refining process, especially vanadium and nickel, which can occur in high concentrations [17, 30]. Some trace elements can cause considerable corrosion to the refining equipments [17, 30]. Mercury may deposit in the equipments, which requires maintenance [60]. Therefore, it is important to determine these trace metal elements in crude oils precisely and accurately to help us decide whether specific metals should be removed prior to refining [61]. On the other hand, on-land surface crude oil or refined oil spills may cause heavy metals to enter soils and water supplies, or regional drainage systems. Some heavy metals are toxic at high doses. For example, Pb can damage nervous systems even at relatively low doses. Additionally, enrichment of heavy metals may occur due to oil spills at sea, which may harm the ecological equilibrium [17]. Therefore, it is becoming more and more important to monitor the potentially hazardous trace elements in crude or refined oils that may do harm to the environment.

The most widely used instrumental analytical techniques for the determination of trace elements in crude oils include atomic absorption spectrometry (AAS) [62], inductively coupled plasma – optical emission spectrometry (ICP-OES) [63], and inductively coupled plasma – mass spectrometry (ICP-MS) [28].

Sample Preparation

Crude oil is a complex mixture of both organic and inorganic matters. Crude oils may be subject to high viscosity, deleterious instrumental effects, and severity of carbon-based analyte interferences [35]. Crude oils with high organic concentration and high viscosity may influence the stability of the plasma or even cause plasma extinction [43]. On the other hand, different phases of sample (i.e., organic vs. inorganic) require different pretreatment for the instruments, or damage of the instruments may occur. Typically aqueous samples are required for ICP-OES and ICP-MS, since high organic load may lead to deposition of carbon on, and clog the nebulizer and plasma tube of the ICP-OES, and the nebulizer, plasma tube, and sampler cone and skimmer cone of the ICP-MS [43]. Therefore, crude oils may not be ideal for direct analysis, and usually require pretreatment prior to being introduced into the analytical instruments mentioned before to obtain high precision and high accuracy results [3]. Four methods have been mainly used in sample preparation of investigating crude oil samples with high organic contents: 1) mineralization by strong acid digestion directly (i.e., wet digestion), 2) mineralization by ashing, or wet or dry combustion, 3) direct introduction of crude oil after dilution with organic solvent, and 4) direct emulsion/micro-emulsion introduction [17].

Direct Introduction Sample Preparation Methods

Crude oils can be introduced into the instrument after diluted with organic solvent in certain appropriate percentages without mineralization [64]. This method is thought to be safer and simpler in that there is no need for crude oil samples to be digested under high pressure and high temperature, whereas the mineralization technique risks venting or even explosion, and sample loss, and is time consuming. However, direct crude oil introduction method is usually subject to some problems that have to be considered. Expensive organometallic standards are necessary, while cost-effective aqueous standards are allowable for mineralization technique and emulsification technique. Solubility issues are encountered by some target elements in crude oils. Additionally, the required large sample dilutions up to ×100, or ×1,000 for heavy crudes will prevent analysts from determining a lot of ultra-trace or trace elements that are a necessity for complete geochemical fingerprinting of crude oils [65]. Furthermore, analysts have to prepare large quantities of expensive organic solvents as rinsing solution, whereas low-cost 2% HNO_3 is enough for rinsing for mineralization and emulsification methods. Long rinsing times imposed by memory effects limit sample throughput [65].

Another method, emulsification, has also been used in trace metal analyses [66, 67]. Crude oil is modified to form an emulsion or micro-emulsion (three component system) with water and a certain kind of surfactant or co-solvent [3]. The surfactant is able to make the system homogeneous and stable for a relatively long time. The organic load and viscosity of the system are reduced significantly because of the homogeneity, stabilization, and dilution of the micro-droplet of oil in water [3]. The emulsion and micro-emulsion could be directly introduced into the instrument for determination. Low-cost aqueous standards can be applied for instruments calibration without carrying out the complicated procedures of sample mineralization. However, it has been reported that emulsions can be stable for only up to 80 hours [68], and as short as 20-50 min [69] for some analytes. This can cause problems when large throughput of oil samples is required. The quantification of a wide range of low abundance elements may be limited by large dilution factors as well.

Mineralization Sample Preparation Methods

The organic structures of oil can be destroyed and eliminated by combustion, or strong acid digestion (i.e., oxidization), thus the oil can be turned to aqueous solutions that have low viscosity and low organic load, which is ideal to be introduced into the analytical instruments. This sample preparation technique is called mineralization [3, 66, 70]. If nearly complete or complete, this can minimize or remove the polyatomic and spectral interferences caused by the complex organic matrix in crude oils, and avoid carbon deposition, which may clog the skimmer or sampler cones of the ICP-MS, and may cause extinction or instability of the plasma [3, 28]. Furthermore, aqueous sample solutions enable the employment of low-cost aqueous standards for instrument calibration [35, 71, 72].

Closed or Open Vessel Digestion

Strong acid digestion methods can be either closed-vessel or open-vessel. Generally, analysts prefer closed-vessel strong acid digestion because it helps avoid loss of volatile species of certain analytes, and potential contamination, and it enables the application of high pressure and high temperature, which attempt to quantify a broad set of low abundance analytes effectively with single analytical methods [25, 73]. The closed-vessel acid digestions can be carried out with a high pressure asher (e.g., HPA-S, Anton Paar, Graz, Austria) [72, 74], or high pressure Parr bombs, or be assisted by a special microwave machine. The high pressure asher digestion systems and microwave-assisted digestion systems share the same operating principle, which is to keep the mixture of reaction reagent and oil sample under the boiling point by increasing the pressure. The microwave-assisted digestion systems use microwave energy to heat, whereas the HPA-S systems use electric energy to heat. The highest digestion pressure and temperature of microwave-assisted digestion systems are lower than HPA-S systems [72, 74]. The microwave-assisted strong acid digestion (~60 minutes) [18] can save a lot of time compared to high pressure asher strong acid digestion (up to 280 minutes including heating and cooling) [72, 74]. High pressure ashers may be subject to contamination problems for certain analytes [74]. HPA-S has proven successful for some studies. However, venting of single reaction vessels during intensive exothermic reactions may happen as volatile oils are decomposed, which may result in sample homogenization problems, setting more limits on sample sizes utilized, or sample contamination or loss. Low abundance analytes determination may be problematic using HPA-S method, as dilutions up to 200 were reported in a literature involving ICP-MS and HPA-S recently [72]. Milestone Inc. developed a new and advanced microwave-assisted strong acid digestion system for crude oil, named SRC (single reaction chamber) microwave system [17, 18]. The difference between the newly introduced SRC microwave system and the traditional rotor based microwave system is discussed below.

Rotor Based or SRC (Single Reaction Chamber) Microwave Digestion

Compared to oven heating, microwave enables the settings of wet digestion processes to be much more flexible in that the pressure and temperature of the improved reactions are able to be monitored and controlled precisely and safely [26, 74], and enhances wet digestion processes for oils considerably [35]. A variety of combinations of digestion running programs and reagents, therefore, can be tested effectively. Additionally, compared to 11 + 0.5 h (heating + cooling) for Parr bomb strong acid digestion, microwave-assisted strong acid digestion saves a lot of time in that it takes only 48 + 22 min for microwave acid digestion to be finished [17]. There are two kinds of microwave digestion systems. One is the rotor based systems with closed single sample digestion vessels, and loaded with multiple vessels. The other is the SRC systems that is equipped with a closed single reaction chamber (SRC) in which multiple sample reaction vessels can be accommodated. Rotor based digestion systems are the most widely used microwave systems, and were introduced in the early 1980's. They have a huge microwave cavity with a rotor inside, which is accommodated with tens of individualized reactors [75]. SRC systems were developed in 2005, characterized by a strengthened single stainless steel chamber [75]. It has some robust advantages described below compared to the more widely used in the early times for fuel oils or natural or refined crude oils, typical and traditional rotor based microwave digestion systems. First of all, in order to prevent combustion-related or exothermic-reaction-related venting and sample loss, a maximum of 0.25 g of sample sizes for rotor based microwave-assisted digestion systems or 0.1 g of sample sizes for Parr bomb strong acid digestion are required, whereas the SRC enables the use of larger sample sizes (e.g., 0.4 g of heavy natural crude oil per 40 ml quartz tube) depending on sample types [17, 26, 74]. For the same dilution, a larger sample size makes the determination of low abundance analytes possible. On the other hand, SRC enables analysts to run multiple samples of different oils at the same time under the same pressure and temperature. Desired MDL (method detection limit), MQL (method quantitation limit), and dilution factors [18] depending on the types of oils (e.g., condensates, light oils, refined oils, fuel oils, or heavy oils with a variety of degrees of homogeneity and analytes abundances) can be obtained by selecting various sizes of digestion vessels within SRC to meet various requirements for oil sample sizes. Additionally, acid boiling is prevented by pre-pressurization of oil samples, and the SRC's inner pressure is identical to that in the sample vessels, which avoids potential sample loss and contamination [17]. What is more, SRC microwave systems specify digestion and seal automatically, which is much faster than the assembly of rotor based microwave systems and single sample Parr bomb method (one sample per bomb).

IDL (Instrumental Detection Limit) is referred to as the smallest mass fraction that is able to be determined for a certain analyte over the background noise within a sample by a certain analytical instrument, which can be achieved by:

IDL (ng/g) = 3 × SD (cps)/Sens. (cps/ng/g) (1) [17],

where SD refers to the standard deviation (1σ) of a number of replicate determinations of the blank for instrument calibration, or 2% (v/v) HNO_3. Counts per second (cps) is the unit. Sens. represents the sensitivity of a certain analytical instrument (e.g., triple-quad ICP-MS, ICP-OES), identical to the value of slope of the instrument calibration curve. MDL (Method Detection Limit) is the smallest mass fraction that can be determined for a certain analyte within a sample by the introduced analytical technique, which can be obtained by:

MDL (ng/g) = IDL × DF = [3 × SD (cps)/Sens. (cps/ng/g)] × DF (2) [17],

where DF refers to the dilution factor. MQL (Method Quantitation Limit) represents the smallest mass fraction that is able to be quantified for a certain analyte using the introduced analytical technique, which can be calculated as:

MQL (ng/g) = 3 × MDL (3) [17].

MDL can be used to estimate a certain technique, and IDL can be used to estimate a particular analytical instrument [17].

Sample Determination Techniques

As the oil samples are pre-treated to have acceptable low viscosity and low organic load, analysts can choose from a variety of techniques to conduct the following determination procedures. Different techniques work in different ways, and have their own characteristics (i.e., advantages and drawbacks) to cater for various purposes and needs. The most commonly used oil sample determination techniques are described below.

Atomic Absorption Spectrometry (AAS)

The most widely used AAS techniques are flame atomic absorption spectrometry (FAAS) and graphite furnace or electrothermal atomic absorption spectrometry (GF-AAS or ET-AAS) [3]. They are able to determine the mass fractions of over 70 different analytes dissolved in liquid samples according to the electronic intensity of absorbed energy of certain wavelengths of lights (usually 190-900 nm) [76, 77]. FAAS includes a flame burner used to atomize sample, a monochromator, and a photon detector [76], and has a sensitivity of a few ppm (parts per million) for most of the analytes, whereas the sensitivity of ET-AAS can be as low as ppb (parts per billion) [76]. Therefore, compared to ET-AAS, much larger sample sizes are required for FAAS to analyze trace or ultra-trace elements [3]. Additionally, AAS can only analyze a particular element per run, while some other instruments (e.g., ICP-OES, ICP-MS) are capable of determining multiple elements (up to over 70) simultaneously [78]. However, FAAS is accompanied by the lowest running and capital costs, and cost per elemental analysis compared to ET-AAS, ICP-OES, and ICP-MS [79]. Jadoon et al., (2016) determined iron, nickel, chromium, and lead in crude oil samples collected from different oil fields of the Kurdistan region, Northern Iraq using FAAS technique [80]. Barbooti (2015) analyzed vanadium, nickel, copper, and iron in crude oil samples of Eastern Baghdad area, Iraq using FAAS [81]. Pessoa et al., (2012) determined calcium, magnesium, and strontium in crude oil samples [82]. Vanadium, nickel, and iron in crude oils and residual fuel oils were analyzed by FAAS for standard test methods ASTM 2005 [3]. Guidroz and Sneddon (2002) analyzed vanadium by FAAS in Venezuelan burned and unburned crude oils [83]. Nickel, cadmium, and lead in unburned and burned Venezuelan crude oils were determined by Hammond et al., (1998) [84]. Platteau and Carrillo (1995) analyzed nickel, iron, and sodium in crude oil by FAAS [85]. Fabec and Ruschak (1985) determined vanadium and nickel in crude oils and heavy crude oil fractions using FAAS [64]. Iron, cadmium, lead, sodium, zinc, copper, and nickel were determined in crude oils using FAAS by Osibanjo et al., (1984) [86]. Nickel alone in fuel oils was analyzed by De la Guardia, M., and Lizondo (1983) using FAAS [87]. Šebor et al., (1982) discussed how the organometallic types of copper and iron influence their determination in oil samples by FAAS [88]. A graphite furnace can be applied to the flame burner to increase the sensitivity [76]. Leite et al., (2015) determined copper, iron, and chromium in used lubricating oil samples using ET-AAS [89]. Luz et al., (2013) analyzed lead, selenium, copper, and cobalt in crude oil, diesel, and gasoline samples by ET-AAS [90]. Arsenic and cadmium were investigated in crudes by de Jesus et al., (2012) using ET-AAS [91]. Iron, nickel, chromium, and vanadium in crude oil samples were analyzed by ET-AAS by Luz and Oliveira (2011) [92]. Dittert et al., (2010) analyzed vanadium and cobalt in crude oil using high-resolution continuum source GF-AAS [93]. Quadros et al., (2010) evaluated Venezuelan and Brazilian crude oils for vanadium and nickel by high-resolution continuum source GF-AAS [94]. Vale et al., (2008) determined non-volatile and volatile vanadium and nickel species in crudes using ET-AAS [10]. Brandão et al., (2006) analyzed nickel in crude oil using GF-AAS [95]. Volatile species of vanadium and nickel were determined by Damin et al., (2005) using GF-AAS [96]. ET-AAS was utilized to analyze vanadium in asphaltenes and diesel samples by Aucélio et al., (2004) [68]. Vanadium again in heavy crude oil samples was determined by Nakamoto et al., (2004) using GF-AAS [97]. Vale et al., (2004) used high-resolution continuum source GF-AAS to analyze nickel alone in crudes [98]. ET-AAS was used by Burguera et al., (2003) to determine chromium in bitumen-in-water emulsion and heavy crudes [99]. Meeravali and Kumar (2001) reported the analysis of vanadium and nickel in naphtha and fuel oil samples using transverse heated ET-AAS [69]. Stigter et al., (2000) determined arsenic, copper, chromium, zinc, and cadmium in crudes by GF-AAS [100]. Chemical modifiers for determination of vanadium in oil and water samples were compared by Thomaidis and Piperaki (1996) utilizing ET-AAS [101]. Heavy crude oil samples were analyzed for vanadium, nickel, mercury, manganese, lead, chromium, cobalt, copper, cadmium, and arsenic by Turunen et al., (1995) using GF-AAS [102]. ET-AAS was conducted by Barrera et al., (1991) for the determination of vanadium in crude oil samples [103]. Alvarado et al., (1990) used GF-AAS to evaluate crude oil samples for their vanadium, nickel, iron, copper, zinc, sodium, manganese, and chromium [104]. Gonzalez et al., (1987) reported to use GF-AAS to determine copper, iron, nickel, lead, and vanadium in petroleum fraction samples [105]. Direct determination by ET-AAS was carried out for nickel alone in gas oils by Bruhn and Cabalín (1983) [106].

Sometimes extremely low abundance elements, such as As, Sb, and Hg, exist in light oils, or condensates, or fuel oils. Even ET-AAS is not able to detect them. At this point, chemical vapor generation atomic absorption spectrometry (CVG-AAS) is proved to have the required sensitivity to solve this problem [66]. However, CVG is subject to high risks of target elements loss because some analyte species are volatile, and tend to adsorb on various surfaces at the level of ppb [107]. Additionally, the application of CVG methods like hydride generation (HG) for antimony and arsenic or cold vapor (CV) for mercury require that the organic structures of oil samples should be destroyed completely. In other words, the oil samples should be mineralized to aqueous phase. By doing this, analysts can reduce or minimize the spectral interferences suffered from by the determination procedures [66]. Trindade et al., (2006) determined arsenic in gasoline using HG-AAS [108]. The CV-AAS was reported in application of analyzing mercury in gasoline by Brandão et al., (2005) [109]. Wilhelm and Bloom (2000) have reviewed and summarized the application of CV-AAS for high sensitivity determination of mercury [110]. De la Guardia et al., (1988) introduced the CVG-AAS technique to investigate the additives of gasoline for manganese and lead [111].

Inductively Coupled Plasma Optical Emission Spectrometry (ICP-OES)

Typically the ICP techniques include inductively coupled plasma optical emission spectrometry (ICP-OES) and inductively coupled plasma mass spectrometry (ICP-MS). The ICP-OES usually has a peristaltic pump, a nebulizer, a spray chamber, a plasma torch, a monochromator, and a detector. ICP instruments are able to conduct multi-element analysis for over 70 analytes per run, and provide a much wider linear dynamic range (up to 4-6 orders of magnitude) than AAS [52]. Commonly, the IDL of ICP-OES ranges from a few to tens of ppm for different analytes, slightly better than FAAS, but less sensitive than GF-AAS and ICP-MS [17, 79]. Therefore, ICP-OES and FAAS are typically used for determination of high abundance elements, such as vanadium, nickel, sulfur, and occasionally others in natural crude oil samples. ICP-OES can be used to do pre-screening to pick out the high abundance elements and low abundance elements in crude oils. The high abundance elements are those whose mass fractions are above the MQL of ICP-OES, and falling into the dynamic range and calibration range of ICP-OES, and thus have a good RSD (relative standard deviation) (usually less than 20%) [17]. The low abundance elements are those below the MQL of ICP-OES, and thus have a poor RSD. Then the low abundance elements need to be further analyzed by more sensitive instruments, such as ICP-MS and GF-AAS. Yang et al., (2017) determined sulfur, vanadium, and nickel in NIST RM (National Institute of Standards and Technology research material) 8505 using ICP-OES [17]. Poirier et al., (2016) applied both ICP-MS and ICP-OES on the determination of iron, nickel, calcium, and vanadium in crudes [112]. Akinlua et al., (2015) used ICP-OES to determine vanadium, nickel, copper, iron, chromium, cobalt, and manganese in crude oils and condensates from South African Basins [113]. Maryutina and Musina (2012) determined aluminum, calcium, manganese, vanadium, silicon, magnesium, nickel, zinc, iron, and barium in heavy oil residues using ICP-AES [114]. Iron, lead, chromium, calcium, boron, copper, aluminum, barium, sulfur, magnesium, molybdenum, nickel, sodium, manganese, silicon, vanadium, potassium, phosphorus, and zinc in crude oils were determined by Nadkarni et al., (2011) using ICP-AES [115]. De Azevedo Mello et al., (2009) analyzed vanadium, sulfur, and nickel in crude oil distillation residues by ICP-OES [116]. Vanadium, nickel, manganese, chromium, cobalt, molybdenum, zinc, cadmium, titanium, and iron were analyzed for crude oils using ICP-OES by de Souza et al., (2006) [117]. ICP-OES was applied to determine vanadium, titanium, chromium, and molybdenum in fuel oil and diesel by de Souza et al., (2006) [118]. Wei et al., (2006) investigated fuel oil by determining vanadium, aluminum, and sodium using ICP-AES [119]. Fischer and Krusberski (2005) determined aluminum, nickel, chromium, copper, lead, and iron in lubricating oil samples by ICP-OES [120]. Determination of refractory elements, chromium, molybdenum, nickel, titanium, and vanadium in used lubricating oils, were conducted by Souza et al., (2004) using ICP-OES [121]. Kuokkanen et al., (2001) analyzed nickel, vanadium, chromium, copper, cadmium, and lead in waste lubricating oil samples by ICP-OES [122]. Lead in gasoline was determined directly by Brenner et al., (1996) using ICP-AES [123]. Murillo and Chirinos (1994) used ICP-AES to determine vanadium, nickel, and sulfur in heavy crude oils [124]. ICP-AES was conducted by Botto (1993) to analyze lead directly in aviation fuel, and aluminum, calcium, iron, sodium, nickel, vanadium, and zinc in NIST SRM 1634b fuel oils [125]. Aluminum, chromium, copper, iron, magnesium, nickel, lead, and sulfur in oil and oil products were determined by Borszéki et al., (1992) using ICP-AES [126].

Inductively Coupled Plasma Mass Spectrometry (ICP-MS)

Typically, the ICP-MS is composed of a peristaltic pump, a nebulizer, a spray chamber, a plasma torch, a mass spectrometer, and a detector. It is another powerful tool in determining trace elements in oil and oil derivatives because of its widely known advantages compared to the others. It has very low LOD (can be as low as 0.02 pg/g, commonly around

1 pg/g) for almost all elements [17]. It is able to do multi-element analysis for a single run, and determine more than 70 analytes. Particularly, it can do isotope related analysis as it distinguish different kinds of ions by mass to charge ratio (m/z). However, similar to ICP-OES, high organic load in the sample solution can make the plasma unstable, or even extinguish. Carbon may deposit on the skimmer cone, or sampler cone, or ion lens, affecting the transport efficiencies of ions. A few years ago, a more advanced technique, triple quadrupole ICP-MS (QQQ-ICP-MS), was introduced, which helped to overcome limitations of the typical single quadrupole ICP-MS (Q-ICP-MS), such as polyatomic interferences, isobaric interferences, and non-mono-charged interferences due to the unique configuration of two quadrupole mass filters on either side of an octopole reaction system (ORS) [17, 72]. Walkner et al., (2017) determined magnesium, aluminum, phosphorus, sulfur, potassium, titanium, vanadium, chromium, manganese, iron, cobalt, nickel, copper, zinc, arsenic, selenium, rubidium, strontium, molybdenum, silver, cadmium, barium, rhenium, lead, and uranium in NIST SRM 1634c fuel oil, NIST SRM 1084a fuel oil, and crude oil samples by QQQ-ICP-MS [72]. Up to 54 trace elements in NIST RM 8505 were analyzed by Yang et al., (2017) using QQQ-ICP-MS [17]. Rare earth elements in crude oil samples were determined by Pereira et al., (2014) using ultrasonic nebulizer ICP-MS [127]. Ortega et al., (2013) analyzed vanadium, nickel, manganese, cobalt, arsenic, molybdenum, barium, and lead in crude oils by ICP-MS [128]. Pereira et al., (2013) determined rare earth elements by ultrasonic nebulizer ICP-MS in crude oil samples [129]. Arsenic and selenium were determined in crude oils by de Albuquerque et al., (2012) using ICP-MS [130]. Pillay et al., (2012) determined 40 trace elements in neem biodiesel and commercial biofuels using high performance ICP-MS [131]. Silver, arsenic, barium, Bismuth, calcium, cadmium, chromium, iron, potassium, magnesium, lithium, manganese, nickel, lead, rubidium, selenium, strontium, thallium, vanadium, and zinc in both light and heavy crude oils were analyzed by Pereira et al., (2010) using ICP-MS [132]. Rare earth elements were determined by Akinlua et al., (2008) in Niger

Delta crude oils using ICP-MS [133]. Saint' Pierre et al., (2006) analyzed silver, cadmium, copper, lead, and thallium in fuel alcohol by electrothermal vaporization ICP-MS [134]. Kowalewska et al., (2005) determined copper in crude oils using ICP-MS [135]. Silver, arsenic, cadmium, copper, cobalt, iron, manganese, nickel, lead, tin, and thallium were determined by Saint' Pierre et al., (2005) using electrothermal vaporization ICP-MS in fuel alcohol [136]. Saint' Pierre et al., (2004) determined cadmium, copper, iron, lead, and thallium in gasoline using electrothermal vaporization ICP-MS [137]. Kelly et al., (2003) analyzed mercury in SRM crude oil samples and refined products by cold vapor ICP-MS [25]. Botto (2002) determined arsenic, sodium, phosphorus, vanadium, lead, and mercury in petroleum naphthas and tars using direct injection ICP-MS [30]. Duyck et al., (2002) analyzed cobalt, yttrium, cadmium, lanthanum, vanadium, barium, molybdenum, and nickel in crude oil and its fractions by ultrasonic nebulization ICP-MS [32]. Copper, manganese, nickel, and tin were analyzed by Saint' Pierre et al., (2002) in gasoline using electrothermal vaporization ICP-MS [138]. Wondimu et al., (2000) determined silver, aluminum, arsenic, barium, Bismuth, calcium, cadmium, cobalt, chromium, copper, iron, mercury, magnesium, molybdenum, nickel, lead, antimony, tin, strontium, titanium, thallium, vanadium, uranium, and zinc in NIST RM 1634b residual fuel oils by ICP-MS [26]. Lead, nickel, and vanadium in Saudi Arabian crude oils were analyzed by Al-Swaidan (1996) using ICP-MS [139]. Al-Swaidan (1994) determined lead and cadmium by ICP-MS in Saudi Arabian petroleum products [140]. Determination of lead and lead isotope ratios was conducted using ICP-MS in gasoline by Lord (1994) [141]. ICP-MS was used to analyze commercial gasolines and crude oil reference materials by Reimer and Miyazaki (1993) for nickel and vanadium [142]. Lord (1991) determined nickel and vanadium in various crude oil samples, vanadium, chromium, manganese, iron, cobalt, nickel, zinc, and lead in NBS 1634b microemulsion, and magnesium, aluminum, titanium, chromium, iron, nickel, copper, molybdenum, silver, tin, and lead in NBS 1085 microemultion [143].

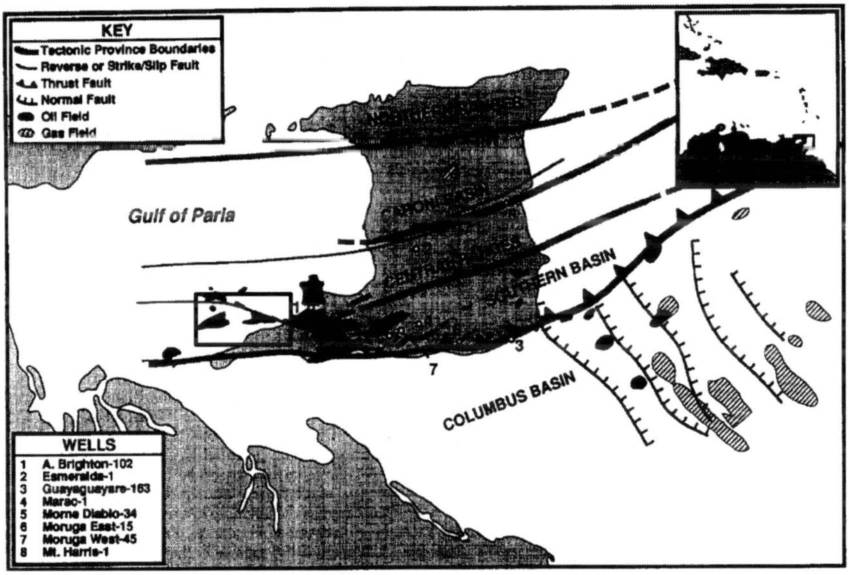

Figure 1. Map of Trinidad showing the locations where the seven crude oil samples from Trinmar Offshore were collected (in the box) (modified from Requejo et al., (1994) [144]).

A Case Study: Natural Crude Oils from Trinidad and NIST RM 8505

The most recently published multi-element analyses for NIST RM 8505 (a high vanadium Venezuelan crude oil, Yang et al., 2017) and crude oil samples from Trinmar Offshore, Trinidad, are summarized in Table 1. Figure 1 shows the locations where the seven crude oil samples from Trinidad were collected.

Trinidad is a small but oil-rich island locating in the northeastern part of the East Venezuela Basin. Due to different stages of structural movement, it can be further divided into five sub basins. However, the southern-most two basins, namely Southern Basin and the Columbus Basin (Figure 1), have attracted most of our attention since they have the highest potential of oil and gas production. According to previous study (Galarraga et al., 2008), the primary source rocks in these areas were in Guayuta

Group, usually including Querecual and San Antonio Formations. Deep water calcareous shales and mudstones, which contain lots of organic compositions and deposited in anoxic environments are dominated in these formations (Casani, 1985). Furthermore, combining with the overlain deltaic depositions, they are undoubtedly served as best reservoirs, with totally more than 2 billion barrels of oil and approximately 20 trillion cubic gas production.

As Table 1 shows, crude oil samples from different places have a huge variety of abundances for a specific element. Different elements in a specific crude oil sample may have a significant difference in abundances. Crude oil samples from the same location may be of great difference for the abundances of a specific element. The abundances can be as low as less than 0.002 ppb for tantalum, and as high as around 27,000,000 ppb for sulfur. The ratio of the highest value and the lowest value can be up to 13,500,000,000. These large spans on abundances of the interested trace elements in crude oils from the same or different origins are most likely to be caused by diverse Eh-pH conditions [54], although how the oxidizing and reducing processes taking place in various pH ambients affect their contents needs further investigation. The precision in terms of RSD (relative standard deviation, 1σ) is less than 4% for most of the elements, although there are still elements not detectable for the ICP techniques. Various classification systems have been reported based on the depositional environments of source rocks using nickel to vanadium ratio and their total contents [44, 54, 55, 145]. Typically, low nickel to vanadium ratio and their high total contents reveal a marine depositional environment, whereas high nickel to vanadium ratio and their low total contents reveal a non-marine or terrigenous depositional environment. Based on this, the nickel to vanadium ratio of NIST RM 8505 is the lowest, and its total content of vanadium and nickel is the highest, so the NIST RM 8505 is most likely to be marine originated. For the seven crude oil samples of Trinidad, only the fourth sample's ratio of nickel and vanadium is over 2, which is much higher than the others, and the fourth sample has the lowest total content of vanadium and nickel, only 22,348 ppb, which is much lower than the others. This indicates an origin closest to the non-marine depositional environment.

Galarraga et al., (2008) [145] did statistical descriptive analyses using vanadium, nickel, and sulfur mass fractions in some nonmature crude oil samples from different Spanish and Venezuelan basins, and utilized the results to complete a generic diagram with four well defined zones by limit vanadium to nickel ratios or nickel contents. This diagram is helpful in the determination of depositional environments of source rocks, and oil-oil, oil-source rock correlations. The data of vanadium and nickel in the seven crude oil samples from Trinidad and NIST RM 8505 were projected on this diagram (Figure 2). It is clear that the source rocks of NIST RM 8505 deposited in a marine anoxic condition, and the source rocks of Trinidad crudes have a terrestrial oxic depositional environment. On the other hand, the nickel to vanadium ratio of NIST RM 8505 is 0.13, which is very close to most of the data of crude oil samples from Urdaneta, Venezuela [145]. This indicates that NIST RM 8505 and the crude oil samples from Urdaneta may have the same origin, and their most likely source rocks are La Luna Formation [145]. The fourth sample of Trinidad and the others may have different origins. However, it may be derived from the same source rocks as the crudes from Ceibita, Venezuela (Figure 2).

REE (rare earth elements), or lanthanides, includes fourteen elements forming a series starting from lanthanum (the lightest) to lutetium (the heaviest). Yttrium is often investigated together with REE due to its similarity in chemical properties to holmium [146]. REE+yttrium (REY) have such similar chemical properties that they usually behave coherently, which leads to predictable distribution patterns when normalized to a standard shale, PAAS (Post Archean Australian Shale). Among the REY elements, cerium and europium are special because they can exist in two oxidation states, while the other elements are typically in only the +3 oxidation state [146]. Cerium can occur in both the +3 and +4 oxidation states. Ce(III) can be partially oxidized to Ce(IV) on the surface of Mn (oxyhydr) oxides under oxic conditions, resulting in its depletion compared to the other REE [147]. Cerium anomalies (Ce_{SN}/Ce_{SN}^*) can be calculated by:

$$Ce_{SN}/Ce_{SN}^* = 2 * [Ce]_{SN}/([La]_{SN} + [Pr]_{SN}) \quad (4) \; [146],$$

where SN stands for shale normalized. Under strongly reducing conditions, europium can exist in its bivalent state. Its anomalies (Eu_{SN}/Eu_{SN}^*) can be calculated by:

$$Eu_{SN}/Eu_{SN}^* = 2 * [Eu]_{SN}/([Sm]_{SN} + [Gd]_{SN}) \quad (5) \; [146].$$

Nd_{SN}/Yb_{SN} can be used to determine if the light REE (LREE) or heavy REE (HREE) are depleted or enriched [148].

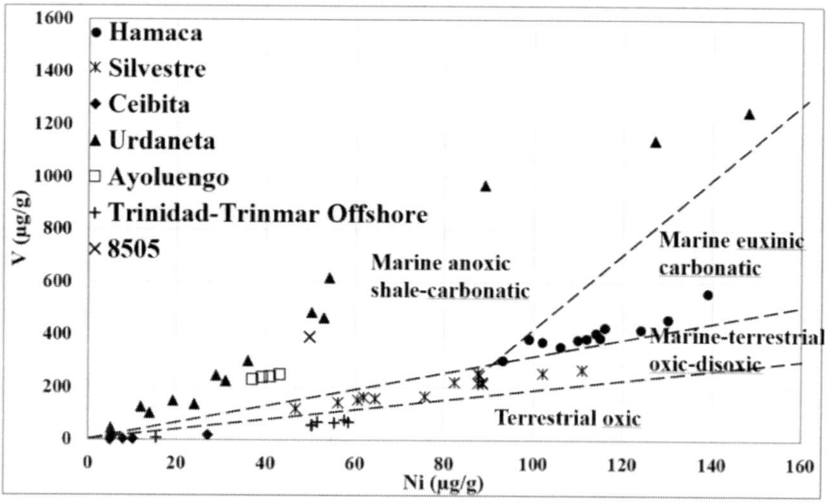

Figure 2. Graphic zonation of vanadium and nickel values as tool for depositional environments interpretations (modified from Galarraga et al., (2008) [145]) (the data of Hamaca, Silvestre, Ceibita, Urdaneta, and Ayoluengo are from Galarraga et al., (2008) [145]).

The ΣREY concentrations in the seven Trinidad crude oils range from 3.6 ppb to 2984 ppb, with an average of 435 ppb. The fourth sample has an extremely high ΣREY concentration compared to the others, 2984 ppb. However, the ΣREY concentrations in the other six Trinidad crude oil samples range from only 3.6 ppb to 16 ppb, with an average of 11 ppb. The Nd_{SN}/Yb_{SN} of the fourth Trinidad crude oil sample is 1.32, indicating a slightly enriched LREE, while the other six Trinidad crude oil samples show a depleted LREE (Nd_{SN}/Yb_{SN} from 0.12 to 0.26, average at 0.19). All the seven Trinidad crude oil samples have non-seawater-like distribution patterns after normalized to the PAAS (Figure 3). The first, fourth, and sixth Trinidad crude oil samples have slightly negative cerium anomalies (Ce_{SN}/Ce_{SN}^* from 0.91 to 0.98, average at 0.95), while the other four samples have small positive cerium anomalies (Ce_{SN}/Ce_{SN}^* from 1.11 to 1.43, average at 1.21). The second, third, sixth, and seventh Trinidad crude oil samples have negative europium anomalies (Eu_{SN}/Eu_{SN}^* from 0.30 to 0.83, average at 0.53), while the first and fourth samples have positive europium anomalies (Eu_{SN}/Eu_{SN}^* are 1.40 and 1.43, respectively, average at 1.42). The fifth sample does not have europium anomalies as its Eu_{SN}/Eu_{SN}^* is 1.00.

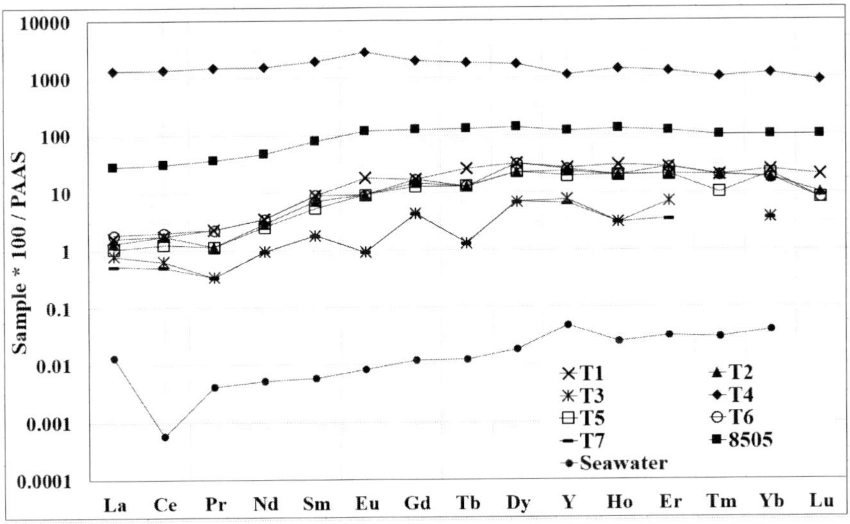

Figure 3. PAAS normalized REY distribution patterns of the seven Trinidad crude oil samples and NIST RM 8505. The REY data of PAAS used are from Nance and Taylor, 1976 [149]. The REY data of seawater used are from Moller et al., (1994) [150], Bau et al., (1995) [151], and James et al., (1995) [152]. Thulium and lutetium were not detectable for the third and seventh Trinidad crude oil samples. Lutetium was not detectable for the seawater sample.

Table 1. Analytical results (mass fractions, ng/g) of the tested elements in the NIST RM 8505 [17] and seven crude oil samples from Trinmar Offshore, Trinidad, by ICP techniques with SRC microwave assisted high pressure and high temperature acid digestion

Elements	T1[a]	T2	T3	T4	T5	T6	T7	8505
Li	1.8	2.5	3.0	2350	4.0	2.3	2.6	2.3
Be	2.9	3.3	2.6	36	4.8	3.2	2.6	2.5
B	109	142	135	7842	349	174	181	107
Na	6095	12777	22677	4266517	18334	8457	25924	13300
Mg	151	480	204	201611	845	377	309	5267
Al	435	560	581	1540381	1415	709	515	668
P	78	64	76	23722	67	79	174	147
K	172	285	255	331395	306	575	324	580
Sc	1.0	1.0	0.2	129	0.9	1.0	0.3	5.2
Ti	1892	1445	524	12869	1422	1905	480	374
Cr	179	229	128	222783	151	134	77	171
Mn	14	8.3	9.9	27053	9.9	20	15	759
Fe	2766	1716	6408	984140	2720	4073	7978	4354
Co	777	485	1836	347	682	886	1924	658
Cu	27	48	20	98756	60	32	27	50
Zn	971	248	681	53726	179	1019	784	416
Ga	203	365	97	399	524	243	103	33.5
Ge	1.2	1.0	2.3	45	0.6	1.4	1.9	2.2
As	117	224	38	994	242	110	34	215
Se	587	520	402	59	631	552	366	371
Rb	0.1	0.2	ND[b]	1437	0.1	0.6	0.05	0.9

Elements	**T1[a]**	**T2**	**T3**	**T4**	**T5**	**T6**	**T7**	**8505**
Sr	11	39	50	218685	40	16	53	288
Y	7.6	6.6	2.1	320	5.5	7.2	1.8	33.1
Zr	43	33	26	762	21	32	14	24.2
Nb	0.5	0.4	0.1	22	0.2	0.2	ND	0.3
Mo	346	722	321	805	722	326	289	264
Ag	0.01	ND	ND	174	ND	ND	ND	0.02
Cd	1.3	24	197	46	25	1.3	175	0.8
Sn	31	36	33	210	34	42	99	81
Sb	2.5	1.9	2.6	442	1.5	1.0	1.5	11.6
Cs	ND	ND	ND	114	ND	0.003	ND	0.08
Ba	80	419	446	14648	35	42	397	438
La	0.6	0.5	0.3	508	0.4	0.7	0.2	10.9
Ce	1.4	1.4	0.5	1105	1.0	1.6	0.4	24.6
Pr	0.2	0.1	0.03	134	0.1	0.2	0.03	3.3
Nd	1.1	0.9	0.3	498	0.8	1.1	0.3	15.2
Sm	0.5	0.4	0.1	108	0.3	0.5	0.1	4.4
Eu	0.2	0.1	0.01	31	0.1	0.1	0.01	1.3
Gd	0.8	0.7	0.2	94	0.6	0.8	0.2	5.9
Tb	0.2	0.1	0.01	14	0.1	0.1	0.01	1
Dy	1.4	1.0	0.3	77	1.0	1.4	0.3	6
Ho	0.3	0.2	0.03	14	0.2	0.2	0.03	1.3
Er	0.8	0.6	0.2	38	0.6	0.8	0.1	3.4
Tm	0.1	0.1	ND	5.2	0.05	0.1	ND	0.5
Yb	0.7	0.5	0.1	33	0.6	0.5	0.1	2.8
Lu	0.1	0.05	ND	4.5	0.04	0.04	ND	0.5

Table 1. (Continued)

Elements	T1[a]	T2	T3	T4	T5	T6	T7	8505
Hf	0.6	0.6	0.5	33	0.3	0.4	0.2	0.5
Ta	0.01	0.002	ND	0.1	0.1	0.1	0.3	0.03
W	2.3	0.6	ND	90	ND	ND	ND	3.1
Tl	0.2	0.1	ND	8.2	ND	ND	ND	0.07
Pb	26	51	140	180128	49	29	120	20.5
Th	0.4	0.4	0.3	262	0.3	0.4	0.2	2.8
U	0.4	0.2	0.2	106	0.05	0.3	0.1	0.4
V	63860	67954	56789	6849	75843	67036	55078	390380
Ni	55332	51533	50398	15499	57544	58582	50158	49809
S	15155778	18912171	13110159	4015333	20692112	15833724	12575015	27051379
V + Ni	119191	119488	107187	22348	133387	125618	105236	440189
Ni / V	0.87	0.76	0.89	2.26	0.76	0.87	0.91	0.13

[a] The first sample from Trinmar Offshore, Trinidad.

[b] Not detectable.

The ΣREY concentration in the NIST RM 8505 is 114 ppb. The Nd_{SN}/Yb_{SN} of the NIST RM 8505 is 0.48, indicating a depleted LREE. The NIST RM 8505 has non-seawater-like distribution patterns after normalized to the PAAS (Figure 3), and slightly negative cerium anomalies (Ce_{SN}/Ce_{SN}^* is 0.94) and positive europium anomalies (Eu_{SN}/Eu_{SN}^* is 1.16).

CONCLUSION

Determination of trace elements in crude oils is of great importance in understanding their formation, oil-oil correlations, oil-source rock correlations, oil migration, maturation of organic matters, and the depositional environment of source rocks. On the other hand, some trace metal elements are problematic because they are not only the poisons to the catalysts, but also can cause significant corrosion to the refining equipment during refinery. Whether specific trace metals should be removed prior to refinery depends on their abundances which are supposed to be determined accurately and precisely. On-land surface crude oil or refined oil spills can result in toxic heavy metal elements entering soils and water bodies, which will threaten the local natural environment.

Crude oils are complex mixtures of both organic and inorganic matters, with high organic load and viscosity, which can cause polyatomic and spectrometric interferences during determination, and instability or even extinction of the plasma. Therefore, pretreatment is required before they are introduced into the analytical instruments. Mineralization by strong acid digestion, mineralization by ashing, or wet or dry combustion, direct introduction of crude oil after dilution with organic solvent, and direct emulsion/micro-emulsion introduction are the most commonly used techniques for sample preparation. The first technique can be conducted with either closed vessel or open vessel. Closed vessel mineralization has been proved to be more efficient in digesting crude oils as much higher temperature and pressure can be applied to the digestion compared to open vessel digestion. For closed vessel digestion, microwave-assisted digestion works better than oven heating digestion in that the pressure and temperature of the improved reactions are able to be monitored and controlled precisely and safely. The rotor based microwave system and SRC (single reaction chamber) microwave system are the most common microwave digestion systems so far. Compared to the typical and traditional rotor based digestion system, the SRC digestion system allows larger sample sizes per sample vessel, which is good for ultra-trace elements determination. Additionally, the SRC systems enable analysts to analyze multiple oil samples at the same time under the same temperature and pressure. Furthermore, the SRC systems avoid potential sample loss and contamination by achieving the same pressure in both the sample vessels and SRC.

The pretreated oil samples are introduced into proper instruments for determination. The most commonly used analytical techniques include FAAS (flame atomic absorption spectrometry), ET-AAS (electrothermal atomic absorption spectrometry) or GF-AAS (graphite furnace atomic absorption spectrometry), CVG-AAS (chemical vapor generation atomic absorption spectrometry), ICP-OES (inductively coupled plasma optical emission spectrometry), and ICP-MS (inductively coupled plasma mass spectrometry). Every technique has its own pros and cons. Generally, FAAS is the cheapest technique, and simplest to be operated, but it has the highest detection limits, and can only analyze a single analyte per run, whereas ET- or GF-AAS is able to determine lower abundance elements due to its much higher sensitivity. CVG-AAS is even more sensitive than ET-AAS. This technique, however, is subject to sample loss as some interested trace elements species are volatile, and oil samples need to be mineralized prior to determination to avoid severe spectral interferences. The ICP techniques are among the most expensive. ICP-OES has a sensitivity a little better than FAAS, while the sensitivity of ICP-MS is the highest. Both ICP-OES and ICP-MS make the multi-element analysis possible (more than 70 elements). However, both ICP techniques do not work well if exposed to high organic load, which may cause instability or even extinction of the plasma. For ICP-MS, carbon can deposit on sampler and skimmer cones and ion lens to affect the transport efficiencies of ions.

REFERENCES

[1] *Wikipedia*. 2017. "Petroleum." Accessed July 21. https://en.wikipedia.org/wiki/Petroleum#cite_note-7.

[2] *EIA Energy Kids – Oil (petroleum)*. 2017. Accessed July 21. https://www.eia.gov/KIDS/energy.cfm?page=oil_home-basics-k.cfm.

[3] Khuhawar, MY; Aslam Mirza, M; Jahangir, TM. "Determination of metal ions in crude oils." In *Crude oil emulsions-Composition stability and characterization*. Intech, 2012.

[4] *Encyclopedia Britannica*. 2017. Accessed July 23. http://www.britannica.com.

[5] Evans, CR; Rogers, MA; Bailey, NJL. "Evolution and alteration of petroleum in western Canada." *Chemical Geology*, 8, no. 3, (1971), 147-170.

[6] Rudzinski, Walter E; Tejraj, M. Aminabhavi. "A review on extraction and identification of crude oil and related products using supercritical fluid technology." *Energy & fuels*, 14, no. 2, (2000), 464-475.

[7] Mansfield, Cliff T; Bhajendra N. Barman; Jane V. Thomas; Anil K. Mehrotra; James M. McCann. "Petroleum and coal." *Analytical Chemistry*, 71, no. 12, (1999), 81-108.

[8] Islas-Flores, CA; Buenrostro-Gonzalez, E; Lira-Galeana, C. "Comparisons between open column chromatography and HPLC SARA fractionations in petroleum." *Energy & fuels*, 19, no. 5, (2005), 2080-2088.

[9] Andersen, Simon Ivar; Andreas Keul; Erling Stenby. "Variation in composition of subfractions of petroleum asphaltenes." *Petroleum Science and Technology*, 15, no. 7-8, (1997), 611-645.

[10] Vale, Maria Goreti R; Marcia M. Silva; Isabel CF Damin; Pedro J. Sanches Filho; Bernhard Welz. "Determination of volatile and non-volatile nickel and vanadium compounds in crude oil using electrothermal atomic absorption spectrometry after oil fractionation into saturates, aromatics, resins and asphaltenes." *Talanta*, 74, no. 5, (2008): 1385-1391.

[11] Seidl, PR; Chrisman, ECAN; Carvalho, CCV; Leal, KZ; De Menezes, SMC. "NMR analysis of asphaltenes separated from vacuum residues by selected solvents." *Journal of dispersion science and technology*, 25, no. 3, (2004), 349-353.

[12] Sharma, BK; Sarowha, SLS; Bhagat, SD; Tiwari, RK; Gupta, SK; Venkataramani, PS. "Hydrocarbon group type analysis of petroleum heavy fractions using the TLC-FID technique." *Fresenius' journal of analytical chemistry*, 360, no. 5, (1998), 539-544.

[13] Seidl, PR; Chrisman, ECAN; Silva, RC; De Menezes, SMC; Teixeira, MAG. "Critical variables for the characterization of asphaltenes extracted from vacuum residues." *Petroleum science and technology*, 22, no. 7-8, (2004), 961-971.

[14] Fan, Tianguang; Jill S. Buckley. "Rapid and accurate SARA analysis of medium gravity crude oils." *Energy & Fuels*, 16, no. 6, (2002), 1571-1575.

[15] Aske, Narve; Harald, Kallevik; Johan, Sjöblom. "Determination of saturate, aromatic, resin, and asphaltenic (SARA) components in crude oils by means of infrared and near-infrared spectroscopy." *Energy & Fuels*, 15, no. 5, (2001), 1304-1312.

[16] Kaminski, Thomas J; Scott Fogler, H; Nick, Wolf; Piyarat, Wattana; Anurag, Mairal. "Classification of asphaltenes via fractionation and the effect of heteroatom content on dissolution kinetics." *Energy & Fuels*, 14, no. 1, (2000), 25-30.

[17] Yang, Weihang; John F. Casey; Yongjun Gao. "A new sample preparation method for crude or fuel oils by mineralization utilizing single reaction chamber microwave for broader multi-element analysis by ICP techniques." *Fuel*, 206, (2017), 64-79.

[18] Casey, John F; Yongjun, Gao; Weihang, Yang; Robert, Thomas. "New Approaches in Sample Preparation and Precise Multielement Analysis of Crude Oils and Refined Petroleum Products Using Single-Reaction-Chamber Microwave Digestion and Triple-Quadrupole ICP-MS." *Spectroscopy*, 31, no. 10, (2016), 11-22.

[19] Curiale, JA. *Distribution and Occurrence of Metals in Heavy Crude Oils and Solid Bitumens--Implications for Petroleum Exploration:*

Section II. Characterization, Maturation, and Degradation. (1987), 207-219.

[20] Alberdi-Genolet, Margarita; Rafael Tocco. "Trace metals and organic geochemistry of the Machiques Member (Aptian–Albian) and La Luna Formation (Cenomanian–Campanian), Venezuela." *Chemical Geology*, 160, no. 1, (1999), 19-38.

[21] Greibrokk, T; Lundanes, E; Norli, HR; Dyrstad, K; Olsen, SD. "Experimental simulation of oil migration—Distribution effects on organic compound groups and on metal/metal ratios." *Chemical Geology*, 116, no. 3-4, (1994), 281-299.

[22] Fonseca, TCO. (2000). PhD Thesis, Chemistry Department, PUC-RJ, Brazil, 135 pp.

[23] Duyck, C. (2001). PhD Thesis, Chemistry Department, PUC-RJ, Brazil, 203p.

[24] López, Liliana; Salvador Lo Mónaco. "Geochemical implications of trace elements and sulfur in the saturate, aromatic and resin fractions of crude oil from the Mara and Mara Oeste fields, Venezuela." *Fuel*, 83, no. 3, (2004), 365-374.

[25] Kelly, W. Robert; Stephen E. Long; Jacqueline L. Mann. "Determination of mercury in SRM crude oils and refined products by isotope dilution cold vapor ICP-MS using closed-system combustion." *Analytical and bioanalytical chemistry*, 376, no. 5, (2003), 753-758.

[26] Wondimu, T; Goessler, W; Irgolic, KJ. "Microwave digestion of "residual fuel oil" (NIST SRM 1634b) for the determination of trace elements by inductively coupled plasma-mass spectrometry." *Fresenius' journal of analytical chemistry*, 367, no. 1, (2000), 35-42.

[27] Bings, Nicolas H. "Direct determination of metals in lubricating oils by laser ablation coupled to inductively coupled plasma time-of-flight mass spectrometry." *Journal of Analytical Atomic Spectrometry*, 17, no. 8, (2002), 759-767.

[28] Lord, Charles J. "Determination of trace metals in crude oil by inductively coupled plasma mass spectrometry with microemulsion

sample introduction." *Analytical chemistry*, 63, no. 15, (1991), 1594-1599.

[29] de Souza, Roseli M; André LS Meliande; Carmem LP da Silveira; Ricardo Q. Aucélio. "Determination of Mo, Zn, Cd, Ti, Ni, V, Fe, Mn, Cr and Co in crude oil using inductively coupled plasma optical emission spectrometry and sample introduction as detergentless microemulsions." *Microchemical Journal*, 82, no. 2, (2006), 137-141.

[30] BOTTO, Robert I. "Trace element analysis of petroleum naphthas and tars using direct injection ICP-MS." *Canadian journal of analytical sciences and spectroscopy*, 47, no. 1, (2002), 1-12.

[31] Kahen, Kaveh; Adelitza, Strubinger; José R. Chirinos; Akbar, Montaser. "Direct injection high efficiency nebulizer-inductively coupled plasma mass spectrometry for analysis of petroleum samples." *Spectrochimica Acta Part B: Atomic Spectroscopy*, 58, no. 3, (2003), 397-413.

[32] Duyck, Christiane; Norbert, Miekeley; Carmem L. Porto da Silveira; Peter Szatmari. "Trace element determination in crude oil and its fractions by inductively coupled plasma mass spectrometry using ultrasonic nebulization of toluene solutions." *Spectrochimica Acta Part B: Atomic Spectroscopy*, 57, no. 12, (2002): 1979-1990.

[33] Kowalewska, Zofia, Anna Ruszczyńska, and Ewa Bulska. "Cu determination in crude oil distillation products by atomic absorption and inductively coupled plasma mass spectrometry after analyte transfer to aqueous solution." *Spectrochimica Acta Part B: Atomic Spectroscopy*, 60, no. 3, (2005), 351-359.

[34] Sneddon, Joseph, Carey Hardaway, Kishore K. Bobbadi, and James N. Beck. "A study of a crude oil spill site for selected metal concentrations remediated by a controlled burning in Southwest Louisiana." *Microchemical journal*, 82, no. 1, (2006): 8-16.

[35] Duyck, Christiane, Norbert Miekeley, Carmem L. Porto da Silveira, Ricardo Q. Aucelio, Reinaldo C. Campos, Patricia Grinberg, and Geisamanda P. Brandao. "The determination of trace elements in crude oil and its heavy fractions by atomic spectrometry."

Spectrochimica Acta Part B: Atomic Spectroscopy, 62, no. 9, (2007), 939-951.

[36] Treibs, A. "Chlorophyll and minerals in organic Häminderivate." *Angew Chemie*, 49, (1936), 682-686.

[37] Baker, Earl W; William Louda, J. "Porphyrin geochemistry of Atlantic Jurassic-Cretaceous black shales." *Organic geochemistry*, 10, no. 4, (1986), 905-914.

[38] Barwise, AJG; Roberts, I. "Diagenetic and catagenetic pathways for porphyrins in sediments." *Organic geochemistry*, 6, (1984), 167-176.

[39] Orr, Wilson L; Kenneth Orris, Emery; John R. Grady. "Preservation of chlorophyll derivatives in sediments off Southern California." *AAPG Bulletin*, 42, no. 5, (1958), 925-962.

[40] Baker, EW; Smith, GD. "Pleistocene changes in chlorophyll pigments." *Advances in organic geochemistry*, (1973), 649-660.

[41] Yen, TF. "Chemical aspects of metals in native petroleum." *Trace Metals in Petroleum*, (1975).

[42] Lewan, Michael Donald. *"Geochemistry of Vanadium and Nickel in Organic Matter of Sedimentary Rocks."* (1981), 0859-0859.

[43] Yang, Weihang. *"High Precision Determination of Trace Elements in Crude Oils by Using Inductively Coupled Plasma-Optical Emission Spectrometry and Inductively Coupled Plasma-Mass Spectrometry."* PhD diss., 2014.

[44] Filby, Royston H. "Origin and nature of trace element species in crude oils, bitumens and kerogens: implications for correlation and other geochemical studies." *Geological Society, London, Special Publications*, 78, no. 1, (1994), 203-219.

[45] Las Ebdon, EHV; Warren, GR; Stephen, JR. "Analysis of geoporphyrins by high temperature inductively coupled plasma mass spectrometry and high performance liquid chromatography inductively coupled plasma mass spectrometry." *J. Anal. At. Spectrom*, 9, (1994), 939-943.

[46] Ball, John S; Wenger, WJ; Hyden, HJ; Horr, CA; Myers, AT. "Metal Content of Twenty-Four Petroleums." *Journal of Chemical and Engineering data*, 5, no. 4, (1960), 553-557.

[47] Al-Shahristani, H; Al-Atyia, MJ. "Vertical migration of oil in Iraqi oil fields: Evidence based on vanadium and nickel concentrations." *Geochimica et Cosmochimica Acta*, 36, no. 9, (1972), 929-938.

[48] El-Gayar, M; Sh, Mostafa, AR; Abdelfattah, AE; Barakat, AO. "Application of geochemical parameters for classification of crude oils from Egypt into source-related types." *Fuel processing technology*, 79, no. 1 (2002), 13-28.

[49] Didyk, Borys M; Yousif IA Alturki; Colin T. Pillinger; Geoffrey Eglinton. "Petroporphyrins as indicators of geothermal maturation." *Nature*, 256, no. 5518, (1975), 563-565.

[50] Barwise, AJG. *Mechanisms involved in altering Deoxophylloerythroetioporphyrin-Etioporphyrin ratios in sediments and oils*. 1987.

[51] Peters, KE; Walters, CC; Moldowan, JM. *The biomarker guide: Volume 2, Biomarkers and isotopes in petroleum systems and earth history*. Cambridge University Press, 2007.

[52] Amorim, Fábio AC; Bernhard, Welz; Antônio CS Costa; Fábio G. Lepri; Maria Goreti R. Vale; Sérgio LC Ferreira. "Determination of vanadium in petroleum and petroleum products using atomic spectrometric techniques." *Talanta*, 72, no. 2, (2007), 349-359.

[53] Choudhary, GG. "Humic substances: Sorptive interactions with environmental chemicals." *Toxicological and Environmental Chemistry*, 6, (1983), 660-672.

[54] Lewan, M. Do. "Factors controlling the proportionality of vanadium to nickel in crude oils." *Geochimica et Cosmochimica Acta*, 48, no. 11, (1984), 2231-2238.

[55] Barwise, AJG. "Role of nickel and vanadium in petroleum classification." *Energy & Fuels*, 4, no. 6, (1990), 647-652.

[56] Ellrich, J; Hirner, A; Stärk, H. "Distribution of trace elements in crude oils from southern Germany." *Chemical Geology*, 48, no. 1-4, (1985), 313-323.

[57] Oluwole, AF; Asubiojo, OI; Nwachukwu, JI; Ojo, JO; Ogunsola, OJ; Adejumo, JA; Filby, RH; Fitzgerald, S; Grimm, C. "Neutron activation analysis of Nigerian crude oils." *Journal of Radioanalytical and Nuclear Chemistry*, 168, no. 1, (1993), 145-152.

[58] Akinlua, A; Ajayi, TR; Adeleke, BB. "Organic and inorganic geochemistry of northwestern Niger Delta oils." *Geochemical journal*, 41, no. 4, (2007), 271-281.

[59] Odermatt, John R; Joseph A. Curiale. "Organically bound metals and biomarkers in the Monterey Formation of the Santa Maria Basin, California." *Chemical geology*, 91, no. 2, (1991), 99-113.

[60] Wilhelm, S. Mark; Lian, Liang; David, Kirchgessner. "Identification and properties of mercury species in crude oil." *Energy & fuels*, 20, no. 1, (2006), 180-186.

[61] Akinlua, A; Torto, N. "Determination of selected metals in Niger Delta oils by graphite furnace atomic absorption spectrometry." *Analytical letters*, 39, no. 9, (2006), 1993-2005.

[62] Langmyhr, FJ; Aadalen, U. "Direct atomic absorption spectrometric determination of copper, nickel and vanadium in coal and petroleum coke." *Analytica Chimica Acta*, 115, (1980), 365-368.

[63] Fabec, Joseph L; Michael L. Ruschak. "Determination of nickel, vanadium, and sulfur in crudes and heavy crude fractions by inductively coupled argon plasma atomic emission spectrometry and flame atomic absorption spectrometry." *Analytical Chemistry*, 57, no. 9, (1985), 1853-1863.

[64] Poirier, Laura; Jenny, Nelson; David, Leong; Lidia, Berhane; Paul, Hajdu; Francisco, Lopez-Linares. "Application of ICP-MS and ICP-OES on the Determination of Nickel, Vanadium, Iron, and Calcium in Petroleum Crude Oils via Direct Dilution." *Energy & Fuels*, 30, no. 5, (2016), 3783-3790.

[65] Ortega, Georgia Sanabria; Christophe, Pécheyran; Gernot, Hudin; Edit, Marosits; Olivier FX, Donard. "Different approaches of crude oil mineralisation for trace metal analysis by ICPMS." *Microchemical Journal*, 106, (2013), 250-254.

[66] Korn, Maria das Graças; Andrade, Denilson Santana; Sodré dos Santos, Bernhard Welz; Maria Goreti Rodrigues Vale; Alete Paixão Teixeira; Daniel de Castro Lima; Sérgio Luis Costa Ferreira. "Atomic spectrometric methods for the determination of metals and

metalloids in automotive fuels–a review." *Talanta*, 73, no. 1, (2007), 1-11.

[67] Pelizzetti, Ezio; Edmondo, Pramauro. "Analytical applications of organized molecular assemblies." *Analytica Chimica Acta*, 169, (1985), 1-29.

[68] Aucélio, Ricardo Q; Adriana, Doyle; Bianca S. Pizzorno; Maria Luiza B. Tristão; Reinaldo C. Campos. "Electrothermal atomic absorption spectrometric method for the determination of vanadium in diesel and asphaltene prepared as detergentless microemulsions." *Microchemical journal*, 78, no. 1, (2004), 21-26.

[69] Meeravali, Noorbasha N; Sunil Jai Kumar. "The utility of a W–Ir permanent chemical modifier for the determination of Ni and V in emulsified fuel oils and naphtha by transverse heated electrothermal atomic absorption spectrometer. This paper is dedicated to the memory of the late Dr S. Gangadharan, Ex-CE, CCCM-BRIT." *Journal of Analytical Atomic Spectrometry*, 16, no. 5, (2001), 527-532.

[70] Mello, Paola A; Juliana SF Pereira; Marcia F. Mesko; Juliano S. Barin; Erico MM Flores. "Sample preparation methods for subsequent determination of metals and non-metals in crude oil—A review." *Analytica chimica acta*, 746, (2012), 15-36.

[71] Sánchez, Raquel; José Luis Todolí; Charles-Philippe Lienemann; Jean-Michel Mermet. "Determination of trace elements in petroleum products by inductively coupled plasma techniques: a critical review." *Spectrochimica Acta Part B: Atomic Spectroscopy*, 88, (2013), 104-126.

[72] Walkner, Christoph; Reinhard, Gratzer; Thomas, Meisel; Syed Nadeem Hussain Bokhari. "Multi-element analysis of crude oils using ICP-QQQ-MS." *Organic Geochemistry*, 103, (2017), 22-30.

[73] Pereira, Juliana SF; Diogo P. Moraes; Fabiane G. Antes; Liange O. Diehl; Maria FP Santos; Regina CL Guimarães; Teresa CO Fonseca; Valderi L. Dressler; Érico MM Flores. "Determination of metals and metalloids in light and heavy crude oil by ICP-MS after digestion by

microwave-induced combustion." *Microchemical Journal*, 96, no. 1, (2010), 4-11.

[74] Wondimu, T; Goessler, W. "Comparison of closed-pressurized and open-refluxed vessel digestion systems for trace elements in the residual fuel oil reference material (SRM 1634c)." *Bulletin of the Chemical Society of Ethiopia*, 14, no. 2. (2000), 99-113.

[75] Milestone Inc. 2017. *Milestone digestion products*. Accessed July 24. http://www.milestonesci.com/digestion.

[76] Labcompare. 2017. *Lab Products*. Accessed August 20. http://www.labcompare.com/Spectroscopy/Atomic-Absorption-Spectro photometer/.

[77] *Wikipedia*. 2017. "Atomic absorption spectroscopy." Accessed August 20. https://en.wikipedia.org/wiki/Atomic_absorption_ spectro scopy.

[78] Perkin Elmer Inc. 2017. *World leader in AA, ICP-OES and ICP-MS, A Guide to Selecting the Appropriate Technique and System*. Accessed August 21. https://www.perkinelmer.com/Content/related materials/brochures/bro_worldleaderaaicpmsicpms.pdf.

[79] Elemental, Thermo. "AAS, GFAAS, ICP or ICP-MS? Which technique should I use." *An elementary overview of elemental analysis*, (2001).

[80] Jadoon, Saima; Abdulfattah Ahmad Amin; Hawbash Khaleel Mahmood; Dalyia Akram Hamoodi; Mohammed Fatih Mohammed Sabir. "Determination of Trace Metals in Crude Oils by Atomic Absorption Spectrophotometry in Khurmala and Guwayar Oil Fields of Kurdistan Region, Iraq." *American Scientific Research Journal for Engineering, Technology, and Sciences (ASRJETS)*, 20, no. 1, (2016), 213-223.

[81] Barbooti, Mahmood M. "Evaluation of Analytical Procedures in the Determination of Trace Metals in Heavy Crude Oils by Flame Atomic Absorption Spectrophotometry." *American Journal of Analytical Chemistry*, 6, no. 04, (2015), 325.

[82] Pessoa, Helen Moura; Rachel Ann Hauser-Davis; Reinaldo Calixto de Campos; Eustáquio Vinicius Ribeiro de Castro; Maria Tereza

Weitzel Dias Carneiro; Geisamanda Pedrini Brandão. "Determination of Ca, Mg, Sr and Ba in crude oil samples by atomic absorption spectrometry." *Journal of Analytical Atomic Spectrometry*, 27, no. 9, (2012), 1568-1573.

[83] Guidroz, Julie M; Joseph Sneddon. "Fate of vanadium determined by nitrous oxide–acetylene flame atomic absorption spectrometry in unburned and burned Venezuelan crude oil." *Microchemical journal*, 73, no. 3, (2002), 363-366.

[84] Hammond, Jenifer L; Yong-Ill Lee; Charles O. Noble; James N. Beck; Edward Proffitt, C; Joseph, Sneddon. "Determination of cadmium, lead, and nickel by simultaneous multielement flame atomic absorption spectrometry in burned and unburned Venezuelan crude oil." *Talanta*, 47, no. 2, (1998), 261-266.

[85] Platteau, Oswald; Marcelo, Carrillo. "Determination of metallic elements in crude oil-water emulsions by flame AAS." *Fuel*, 74, no. 5, (1995), 761-767.

[86] Osibanjo, Oladele; Samuel E. Kakulu; Sunday O. Ajayi. "Analytical application of inorganic salt standards and mixed-solvent systems to trace-metal determination in petroleum crudes by atomic-absorption spectrophotometry." *Analyst*, 109, no. 2, (1984), 127-129.

[87] De la Guardia, M; Lizondo, MJ. "Direct determination of nickel in fuel oil by atomic absorption spectrometry using emulsions." *Atomic spectroscopy*, 4, no. 6, (1983), 208-211.

[88] Šebor, G; Lang, I; Kolihova, D; Weisser, O. "Effect of the type of organometallic iron and copper compounds on the determination of both metals in petroleum samples by flame atomic-absorption spectroscopy." *Analyst*, 107, no. 1280, (1982), 1350-1355.

[89] Leite, Clarice C; Ariane V. Zmozinski; Maria Goreti R. Vale; Márcia M. Silva. "Determination of Fe, Cr and Cu in used lubricating oils by ET AAS using a microemulsion process for sample preparation." *Analytical Methods*, 7, no. 8, (2015), 3363-3371.

[90] Luz, Maciel S; Angerson N. Nascimento; Pedro V. Oliveira. "Fast emulsion-based method for simultaneous determination of Co, Cu,

Pb and Se in crude oil, gasoline and diesel by graphite furnace atomic absorption spectrometry." *Talanta*, 115, (2013), 409-413.

[91] de Jesus, Alexandre; Ariane Vanessa Zmozinski; Isabel Cristina Ferreira Damin; Márcia Messias Silva; Maria Goreti Rodrigues Vale. "Determination of arsenic and cadmium in crude oil by direct sampling graphite furnace atomic absorption spectrometry." *Spectrochimica Acta Part B: Atomic Spectroscopy*, 71, (2012), 86-91.

[92] Luz, Maciel Santos; Pedro Vitoriano Oliveira. "Simultaneous determination of Cr, Fe, Ni and V in crude oil by emulsion sampling graphite furnace atomic absorption spectrometry." *Analytical Methods*, 3, no. 6, (2011), 1280-1283.

[93] Dittert, Ingrid M; Jessee SA Silva; Rennan GO Araujo; Adilson J. Curtius; Bernhard Welz; Helmut Becker-Ross. "Simultaneous determination of cobalt and vanadium in undiluted crude oil using high-resolution continuum source graphite furnace atomic absorption spectrometry." *Journal of Analytical Atomic Spectrometry*, 25, no. 4, (2010), 590-595.

[94] Quadros, Daiane PC; Eduardo S. Chaves; Fábio G. Lepri; Daniel LG Borges; Bernhard Welz; Helmut Becker-Ross; Adilson J. Curtius. "Evaluation of Brazilian and Venezuelan crude oil samples by means of the simultaneous determination of Ni and V as their total and non-volatile fractions using high-resolution continuum source graphite furnace atomic absorption spectrometry." *Energy & Fuels*, 24, no. 11, (2010), 5907-5911.

[95] Brandão, Geisamanda Pedrini; Reinaldo Calixto de Campos; Eustáquio Vinicius Ribeiro de Castro; Honério Coutinho de Jesus. "Direct determination of nickel in petroleum by solid sampling–graphite furnace atomic absorption spectrometry." *Analytical and bioanalytical chemistry*, 386, no. 7-8, (2006), 2249-2253.

[96] Damin, Isabel CF; Maria Goreti R. Vale; Marcia M. Silva; Bernhard Welz; Fabio G. Lepri; Walter NL dos Santos; Sérgio LC Ferreira. "Palladium as chemical modifier for the stabilization of volatile nickel and vanadium compounds in crude oil using graphite furnace

atomic absorption spectrometry." *Journal of Analytical Atomic Spectrometry*, 20, no. 12, (2005), 1332-1336.

[97] Nakamoto, Yoshikazu; Tatsuro Ishimaru; Nobutaka Endo; Koji MATSUSAKI. "Determination of vanadium in heavy oils by atomic absorption spectrometry using a graphite furnace coated with tungsten." *Analytical Sciences*, 20, no. 4, (2004), 739-741.

[98] Vale, Maria Goreti R; Isabel CF Damin; Aline Klassen; Márcia M. Silva; Bernhard Welz; Alessandra F. Silva; Fábio G. Lepri; Daniel LG Borges; Uwe Heitmann. "Method development for the determination of nickel in petroleum using line-source and high-resolution continuum-source graphite furnace atomic absorption spectrometry." *Microchemical Journal*, 77, no. 2, (2004), 131-140.

[99] Burguera, José L; Rita M. Avila-Gómez; Marcela Burguera; Raquel Antón de Salager; Jean-Louis Salager; Carlos L. Bracho; Margarita Burguera-Pascu et al., "Optimum phase-behavior formulation of surfactant/oil/water systems for the determination of chromium in heavy crude oil and in bitumen-in-water emulsion." *Talanta*, 61, no. 3, (2003), 353-361.

[100] Stigter, JB; De Haan, HPM; Guicherit, R; Dekkers, CPA; Daane, ML. "Determination of cadmium, zinc, copper, chromium and arsenic in crude oil cargoes." *Environmental Pollution*, 107, no. 3, (2000), 451-464.

[101] Thomaidis, Nikolaos S; Efrosini A. Piperaki. "Comparison of chemical modifiers for the determination of vanadium in water and oil samples by electrothermal atomization atomic absorption spectrometry." *Analyst*, 121, no. 2, (1996), 111-117.

[102] Turunen, Minna; Sirpa Peräniemi; Markku Ahlgrén; Henrik Westerholm. "Determination of trace elements in heavy oil samples by graphite furnace and cold vapour atomic absorption spectrometry after acid digestion." *Analytica chimica acta*, 311, no. 1, (1995), 85-91.

[103] Barrera, P; Bermejo, C. Pita Calvo; Bermejo Martinez, YF. "Sample preparation procedures for vanadium determination in petroleum by

atomic absorption spectrometry with electrothermal atomization." *Analytical letters*, 24, no. 3, (1991), 447-458.

[104] Alvarado, José; Ana R. Picón; Manganaro de Vecchi, C. "Microwave wet-acid digestion in the preparation of crude oil samples for the AAS determination of their chromium, copper, iron, manganese, sodium, nickel, vanadium and zinc content." *Acta Chim. Ven*, 41, (1990), 306-319.

[105] Gonzalez, MC; Rodriguez, AR; Gonzalez, V. "Determination of vanadium, nickel, iron, copper, and lead in petroleum fractions by atomic absorption spectrophotometry with a graphite furnace." *Microchemical journal*, 35, no. 1, (1987), 94-106.

[106] Bruhn, Carlos F; Vivianne G. Cabalín. "Direct determination of nickel in gas oil by atomic absorption spectrometry with electrothermal atomization." *Analytica Chimica Acta*, 147, (1983), 193-203.

[107] Maier, EA; Griepink, B; Ph Quevauviller, eds. *Quality assurance for environmental analysis: method evaluation within the measurements and testing programme (BCR)*. Vol. 17. Elsevier, 1995.

[108] Trindade, Jemmla Meira; Aldaléa Lopes Marques; Gisele Simone Lopes; Edmar Pereira Marques; Jiujun Zhang. "Arsenic determination in gasoline by hydride generation atomic absorption spectroscopy combined with a factorial experimental design approach." *Fuel*, 85, no. 14, (2006), 2155-2161.

[109] Brandão, Geisamanda Pedrini; Reinaldo Calixto de Campos; and Aderval Severino Luna. "Determination of mercury in gasoline by cold vapor atomic absorption spectrometry with direct reduction in microemulsion media." *Spectrochimica Acta Part B: Atomic Spectroscopy*, 60, no. 5, (2005), 625-631.

[110] Wilhelm, S. Mark; Nicolas Bloom. "Mercury in petroleum." *Fuel Processing Technology*, 63, no. 1, (2000), 1-27.

[111] De la Guardia, Miguel; Adela R. Maurí; Carlos Mongay. "Atomic absorption spectrometric determination of gasoline additives by vapour phase sample introduction." *Journal of Analytical Atomic Spectrometry*, 3, no. 7, (1988), 1035-1038.

[112] Poirier, Laura; Jenny Nelson; David Leong; Lidia Berhane; Paul Hajdu; Francisco Lopez-Linares. "Application of ICP-MS and ICP-OES on the Determination of Nickel, Vanadium, Iron, and Calcium in Petroleum Crude Oils via Direct Dilution." *Energy & Fuels*, 30, no. 5, (2016), 3783-3790.

[113] Akinlua, A; Sigedle, A; Buthelezi, T; Fadipe, OA. "Trace element geochemistry of crude oils and condensates from South African Basins." *Marine and Petroleum Geology*, 59, (2015), 286-293.

[114] Maryutina, TA; Musina, NS. "Determination of metals in heavy oil residues by inductively coupled plasma atomic emission spectroscopy." *Journal of analytical chemistry*, 67, no. 10, (2012), 862-867.

[115] Nadkarni, RA Kishore; J. David Hwang; Lyman Young. "Multielement Analysis of Crude Oils Using Inductively Coupled Plasma-Atomic Emission Spectrometry." *Journal of ASTM International*, 8, no. 10, (2011), 1-11.

[116] de Azevedo Mello; Paola, Juliana Severo Fagundes Pereira; Diogo Pompéu de Moraes; Valderi Luiz Dressler; Érico Marlon de Moraes Flores; Guenter Knapp. "Nickel, vanadium and sulfur determination by inductively coupled plasma optical emission spectrometry in crude oil distillation residues after microwave-induced combustion." *Journal of analytical atomic spectrometry*, 24, no. 7, (2009), 911-916.

[117] de Souza, Roseli M; André LS Meliande; Carmem LP da Silveira; Ricardo Q. Aucélio. "Determination of Mo, Zn, Cd, Ti, Ni, V, Fe, Mn, Cr and Co in crude oil using inductively coupled plasma optical emission spectrometry and sample introduction as detergentless microemulsions." *Microchemical Journal*, 82, no. 2, (2006), 137-141.

[118] de Souza, Roseli M; Bárbara M. Mathias; Ieda S. Scarminio; Carmem Lúcia P. da Silveira; Ricardo Q. Aucélio. "Comparison between two sample emulsification procedures for the determination of Mo, Cr, V and Ti in diesel and fuel oil by ICP OES along with factorial design." *Microchimica Acta*, 153, no. 3, (2006), 219-225.

[119] Wei, HJ; Guan, DL; Sun, PT; Wang, HZ; Wang, ZE; Luo, ZP; Li, SH. "Study on determination of sodium, vanadium and aluminum contents in fuel oil by ICP-AES." *Guang pu xue yu guang pu fen xi= Guang pu*, 26, no. 2, (2006), 340-343.
[120] Fischer, Johann L; Nicolle B. Krusberski. "An evaluation of emulsions in wear-metal-in-oil analyses." *South African Journal of Chemistry*, 58, (2005), 131-137.
[121] Souza, Roseli Martins; Carmem Lúcia P. da SILVEIRA; Ricardo Queiroz Aucelio. "Determination of refractory elements in used lubricating oil by ICPOES employing emulsified sample introduction and calibration with inorganic standards." *Analytical sciences*, 20, no. 2, (2004), 351-355.
[122] Kuokkanen, Toivo; Paavo Perämäki; Ilkka Välimäki; Hannu Rönkkömäki. "Determination of heavy metals in waste lubricating oils by inductively coupled plasma–optical emission spectrometry." *International journal of environmental analytical chemistry*, 81, no. 2, (2001), 89-100.
[123] Brenner, IB; Zander, A; Kim, S; Shkolnik, J. "Direct determination of lead in gasoline using emulsification and argon and argon–oxygen inductively coupled plasma atomic emission spectrometry." *Journal of Analytical Atomic Spectrometry*, 11, no. 2, (1996), 91-97.
[124] Murillo, Miguel; Jose Chirinos. "Use of emulsion systems for the determination of sulfur, nickel and vanadium in heavy crude oil samples by inductively coupled plasma atomic emission spectrometry." *Journal of Analytical Atomic Spectrometry*, 9, no. 3, (1994), 237-240.
[125] Botto, Robert I. "Applications of ultrasonic nebulization in the analysis of petroleum and petrochemicals by inductively coupled plasma atomic emission spectrometry." *Journal of Analytical Atomic Spectrometry*, 8, no. 1, (1993), 51-57.
[126] Borszéki, János; Günter Knapp; Pál Halmos; László Bartha. "Sample preparation procedure for the determination of sulphur and trace metals in oil products by the ICP with a minitorch using emulsions." *Microchimica Acta*, 108, no. 3, (1992), 157-161.

[127] Pereira, JSF; Pereira, LSF; Mello, PA; Guimaraes, RCL; Guarnieri, RA; Fonseca, TCO; Flores, EMM. "Microwave-induced combustion of crude oil for further rare earth elements determination by USN–ICP-MS." *Analytica chimica acta*, 844, (2014), 8-14.

[128] Ortega, Georgia Sanabria; Christophe Pécheyran; Gernot Hudin; Edit Marosits; Olivier FX Donard. "Different approaches of crude oil mineralisation for trace metal analysis by ICPMS." *Microchemical Journal*, 106, (2013), 250-254.

[129] Pereira, JSF; Picoloto, RS; Pereira, LSF; Guimarães, RCL; Guarnieri, RA; Flores, EMM. "High-efficiency microwave-assisted digestion combined to in situ ultraviolet radiation for the determination of rare earth elements by ultrasonic nebulization ICPMS in crude oils." *Analytical chemistry*, 85, no. 22, (2013), 11034-11040.

[130] de Albuquerque; Fernanda Inda; Christiane B. Duyck; Teresa Cristina O. Fonseca; Tatiana D. Saint'Pierre. "Determination of As and Se in crude oil diluted in xylene by inductively coupled plasma mass spectrometry using a dynamic reaction cell for interference correction on 80 Se." *Spectrochimica Acta Part B: Atomic Spectroscopy*, 71, (2012), 112-116.

[131] Pillay, AE; Elkadi, M; Fok, SC; Stephen, S; Manuel, J; Khan, MZ; Unnithan, S. "A comparison of trace metal profiles of neem biodiesel and commercial biofuels using high performance ICP-MS." *Fuel*, 97, (2012), 385-389.

[132] Pereira, Juliana SF; Diogo P. Moraes; Fabiane G. Antes; Liange O. Diehl; Maria FP Santos; Regina CL Guimarães; Teresa CO Fonseca; Valderi L. Dressler; Érico MM Flores. "Determination of metals and metalloids in light and heavy crude oil by ICP-MS after digestion by microwave-induced combustion." *Microchemical Journal*, 96, no. 1, (2010), 4-11.

[133] Akinlua, A; Torto, N; Ajayi, TR. "Determination of rare earth elements in Niger Delta crude oils by inductively coupled plasma-mass spectrometry." *Fuel*, 87, no. 8, (2008), 1469-1477.

[134] Saint'Pierre, Tatiana Dillenburg; Vera Lúcia Azzolin Frescura; Adilson José Curtius. "The development of a method for the

determination of trace elements in fuel alcohol by ETV-ICP-MS using isotope dilution calibration." *Talanta*, 68, no. 3, (2006), 957-962.

[135] Kowalewska, Zofia; Anna Ruszczyńska; Ewa Bulska. "Cu determination in crude oil distillation products by atomic absorption and inductively coupled plasma mass spectrometry after analyte transfer to aqueous solution." *Spectrochimica Acta Part B: Atomic Spectroscopy*, 60, no. 3, (2005), 351-359.

[136] Saint'Pierre, Tatiana Dillenburg; Tatiane de Andrade Maranhão; Vera Lúcia Azzolin Frescura; Adilson José Curtius. "The development of a method for the determination of trace elements in fuel alcohol by electrothermal vaporization–inductively coupled plasma mass spectrometry using external calibration." *Spectrochimica Acta Part B: Atomic Spectroscopy*, 60, no. 5, (2005), 605-613.

[137] Saint'Pierre, Tatiana D; Lúcia F. Dias; Sandra M. Maia; Adilson J. Curtius. "Determination of Cd, Cu, Fe, Pb and Tl in gasoline as emulsion by electrothermal vaporization inductively coupled plasma mass spectrometry with analyte addition and isotope dilution calibration techniques." *Spectrochimica Acta Part B: Atomic Spectroscopy*, 59, no. 4, (2004), 551-558.

[138] Saint'Pierre, Tatiana D; Lúcia Felicidade Dias; Dirce Pozebon; Ricardo Q. Aucélio; Adilson J. Curtius; Bernhard Welz. "Determination of Cu, Mn, Ni and Sn in gasoline by electrothermal vaporization inductively coupled plasma mass spectrometry, and emulsion sample introduction." *Spectrochimica Acta Part B: Atomic Spectroscopy*, 57, no. 12, (2002), 1991-2001.

[139] Al-Swaidan, Hassan M. "The determination of lead, nickel and vanadium in Saudi Arabian crude oil by sequential injection analysis/inductively-coupled plasma mass spectrometry." *Talanta*, 43, no. 8, (1996), 1313-1319.

[140] Al-Swaidan, HM. "Microemulsion determination of lead and cadmium in Saudi Arabian petroleum products by inductively coupled plasma mass spectrometry (ICP/MS)." *Science of the total environment*, 145, no. 1-2, (1994), 157-161.

[141] Lord, Charles J. "Determination of lead and lead isotope ratios in gasoline by inductively coupled plasma mass spectrometry." *Journal of Analytical Atomic Spectrometry*, 9, no. 5, (1994), 599-603.

[142] Reimer, Raul A; Akira Miyazaki. "Determination of nickel and vanadium in commercial gasolines and crude oil reference materials by inductively coupled plasma mass spectrometry with micro-emulsion sample introduction." *Analytical sciences*, 9, no. 1, (1993), 157-159.

[143] Lord, Charles J. "Determination of trace metals in crude oil by inductively coupled plasma mass spectrometry with microemulsion sample introduction." *Analytical chemistry*, 63, no. 15, (1991), 1594-1599.

[144] Requejo, AG; Wielchowsky, CC; Klosterman, MJ; Sassen, R. "Geochemical characterization of lithofacies and organic facies in Cretaceous organic-rich rocks from Trinidad, East Venezuela Basin." *Organic Geochemistry*, 22, no. 3-5, (1994), 441IN1-459IN2.

[145] Galarraga, F; Reategui, K; Martïnez, A; Martínez, M; Llamas, JF; Márquez, G. "V/Ni ratio as a parameter in palaeoenvironmental characterisation of nonmature medium-crude oils from several Latin American basins." *Journal of Petroleum Science and Engineering*, 61, no. 1, (2008), 9-14.

[146] Tostevin, Rosalie; Graham A. Shields; Gary M. Tarbuck; Tianchen He; Matthew O. Clarkson; Rachel A. Wood. "Effective use of cerium anomalies as a redox proxy in carbonate-dominated marine settings." *Chemical Geology*, 438, (2016), 146-162.

[147] German, Christopher R; Henry Elderfield. "Application of the Ce anomaly as a paleoredox indicator: the ground rules." *Paleoceanography*, 5, no. 5, (1990), 823-833.

[148] Jiang, Lei; Chunfang, Cai; Richard H. Worden; Kaikai, Li; Lei, Xiang; Xuelei, Chu; Anjiang, Shen; Wenjun, Li. "Rare earth element and yttrium (REY) geochemistry in carbonate reservoirs during deep burial diagenesis: Implications for REY mobility during thermochemical sulfate reduction." *Chemical Geology*, 415, (2015), 87-101.

[149] Nance, WB; Taylor, SR. "Rare earth element patterns and crustal evolution—I. Australian post-Archean sedimentary rocks." *Geochimica et Cosmochimica Acta*, 40, no. 12, (1976), 1539-1551.

[150] Moller, P; Dulski, P; Bau, M. "Rare-earth element adsorption in a seawater profile above the East Pacific Rise." *Chemie Der Erde-Geochemistry*, 54, no. 2, (1994), 129-149.

[151] Bau, M; Dulski, PETER; Moller, P. "Yttrium and holmium in South Pacific seawater: vertical distribution and possible fractionation mechanisms." *Oceanographic Literature Review*, 11, no. 42, (1995), 955.

[152] James, RH; Elderfield, H; Palmer, MR. "The chemistry of hydrothermal fluids from the Broken Spur site, 29 N Mid-Atlantic Ridge." *Geochimica et Cosmochimica Acta*, 59, no. 4, (1995), 651-659.

In: Solution Chemistry
Editor: Yongliang Xiong

ISBN: 978-1-53613-101-7

Chapter 7

SOLUTION CHEMISTRY OF TRANSITION METAL COMPLEXES: A CASE STUDY OF VANADIUM

Daniele Sanna*
Institute of Biomolecular Chemistry,
National Research Council, Trav. La Crucca 3,
Reg. Baldinca, 07040 Li Punti, Sassari, Italy

ABSTRACT

Solution chemistry of transition metal complexes can be very complicated and difficult to study. The establishment of equilibria depends on the nature of the specific metal ion, its kinetic lability, the thermodynamic stability of the complexes, its oxidation state, the coordinating capability of the solvent where the complexes are dissolved, etc. In fact, while solid complexes represent a "frozen image" of a changing situation, solution chemistry represents a mutating system which can be influenced by a series of factors. A case study to describe the solution chemistry of transition metal ions is represented by vanadium. The solution chemistry of its most important oxidation states, +IV and +V, has been widely investigated because two specific

* Corresponding Author Email: Daniele.Sanna@icb.cnr.it.

spectroscopic techniques are available: ^{51}V NMR (Nuclear Magnetic Resonance) for V(V) and EPR (Electron Paramagnetic Resonance or ESR, Electron Spin Resonance) for V(IV).

In fact, V(V) has no unpaired electrons and can be studied with NMR, while V(IV) has a $3d^1$ electronic configuration which makes it suitable for EPR spectroscopy. These two techniques permitted the detailed investigation of the solution chemistry of these two oxidation states of the metal and the obtained information is complementary because the two techniques have different limitations. When a specific vanadium complex is dissolved in solution different equiibria can take place: i) redox reactions with interconversion between the two oxidation states; ii) ligand exchange, depending on the thermodynamic stability of the complexes; iii) hydrolysis reactions, with eventual polymerization and/or precipitation of hydroxydes; iv) equilibria between different geometries of the complexes; v) eventual exchange reactions with coordinating solvents.

Solution chemistry of vanadium complexes will be discussed in comparison with other transition metal ions. The conclusions drawn for vanadium complexes will be extended to other metal ions which have similar features, while for those which significantly differ from it, the possible scenarios will be depicted.

Keywords: vanadium, transition metals, NMR, EPR, UV-Vis spectroscopy

INTRODUCTION

IUPAC (International Union of Pure and Applied Chemistry) defines a transition element as "an element whose atom has an incomplete d sub-shell, or which can give rise to cations with an incomplete d sub-shell" [1].

The binding in transition metal complexes is usually explained, in simple terms, with crystal field theory [2]. This theory considers the donor atoms in metal complexes as punctiform charges; the interaction between these charges and the electrons in the outer level of the metal causes a separation of the d-orbitals in two or more energetic levels (Scheme 1). The energy difference between these levels is such that these species absorb in the visible range of the electromagnetic spectrum, and therefore transition metal complexes are very often colored species. The most common geometries for coordination numbers 4 and 6 are tetrahedral and octahedral, respectively. For coordination number 5 square pyramidal and trigonal bipyramid geometries are possible (Scheme 1).

Another theory which can be used to explain the affinities of various donor atoms for different metal ions is the "hard" and "soft" acids and bases theory [3]. Chemical species which are small, have high charge and are weakly polarizable are defined as "hard"; "soft" species are those which are big, have low charge and are strongly polarizable. Therefore, for example V ions, V^{4+} and V^{5+}, which are small, have high charges and are weakly polarizable, are considered hard Lewis acids and have high affinities for "hard" Lewis bases like negatively charged oxygen donors. The affinity for oxygen is such that usually V in these two oxidation states does not exist, except few examples (see below), as a "naked" cations but more commonly as $V(IV)O^{2+}$ and $V(V)O_2^+$.

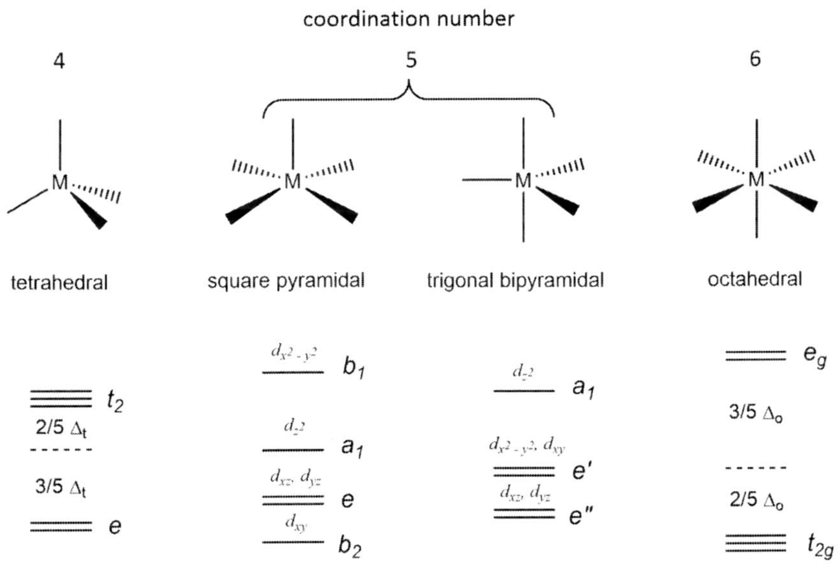

Scheme 1. Common geometries for coordination numbers 4, 5 and 6 found in transition metal complexes and energy of d orbitals. t_2 and t_{2g} represent d_{xy}, d_{xz} and d_{yz} orbitals while e and e_g include d_{z^2} and $d_{x^2-y^2}$.

Considering the thermodynamic stability, the complexes can be more or less stable depending on the specific metal ion, its oxidation state, the involved donor atoms, the coordinating capability of the solvent, etc. Thermodynamic stability is not related to the rate at which the complexes exchange the coordinated ligands. For example, the rate exchange of coordinated water molecules of aqua (aqua is the name of H_2O when it coordinates in metal complexes) transition metal ions encompasses almost twenty orders of magnitude, largely as a consequence of differing d orbital electronic occupancies, which influence transition state energetics and stereochemistries [4].

For example $[Ir(H_2O)_6]^{3+}$ exchanges coordinated water molecule extremely slowly with a water exchange on each Ir^{3+} every ~50 years, while $[Eu(H_2O)_7]^{2+}$ has an extremely fast exchange with a water exchange event every ~2.8×10^{-11} s.

Therefore the rate of exchange of coordinated donors can be high in a thermodynamically stable complex, while can be very low in an unstable complex. The existence of a certain species in solution does not necessarily mean that this species is particularly stable, but this could be a thermodynamically unstable species and its presence is simply due to its kinetic inertness.

Therefore thermodynamic stability is not related to the kinetic lability, but unstable complexes will at the end be converted to something more stable, more or less quickly depending on their kinetic lability.

A Case Study: Vanadium

Vanadium (V) has different oxidation states ranging from +II to +V, but the most important ones, those which will be described in this chapter, are +IV and +V. V(IV) has a $[Ar]3d^1$ electronic configuration and therefore, having one unpaired electron in the 3d orbitals, is suitable for being studied with Electron Spin Resonance (ESR, or EPR, Electron Paramagnetic Resonance) spectroscopy. V(V) having no unpaired electrons can be studied with ^{51}V NMR spectroscopy. Both these techniques take advantage of the isotopic distribution of this metal, in fact 99.75% of it is ^{51}V, while ^{50}V constitutes only 0.25% and can be neglected.

The isotope ^{51}V has a I (nuclear spin) = 7/2 and therefore, the interaction of the unpaired electron with this nucleus originates (2*n*I + 1 =) 8 lines.

In both these oxidation states V has a high charge/radius ratio and therefore it rarely exists as V^{5+} and V^{4+}; as already mentioned, the most frequent forms of these oxidation states are oxidovanadium ions, that is $V(IV)O^{2+}$ (Scheme 2) and $V(V)O_2^+$, respectively. This is not a particular feature of this element but it is common for all transition metal ions having a high charge/radius ratio; among the elements of the first row of transition metals examples are $Ti(IV)O^{2+}$ and $Cr(V)O^{3+}$.

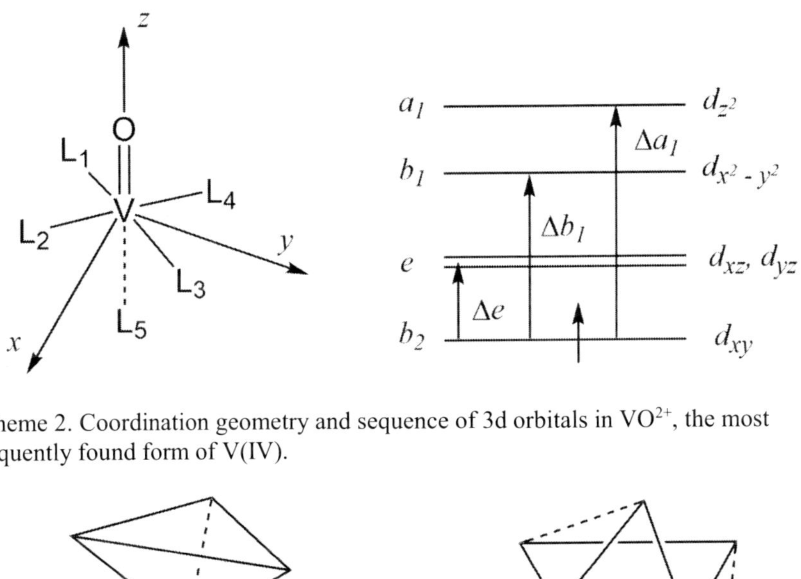

Scheme 2. Coordination geometry and sequence of 3d orbitals in VO^{2+}, the most frequently found form of V(IV).

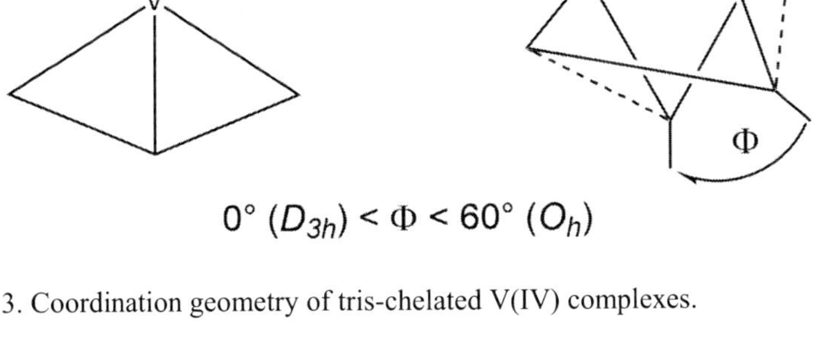

Scheme 3. Coordination geometry of tris-chelated V(IV) complexes.

Tris-chelated complexes formed by bidentate ligands with V(IV), where the oxido ligand has been lost, only exist with particular molecules, for example catecholates. In these cases the coordination geometry is intermediate between the octahedral and the trigonal prismatic (Scheme 3).

^{51}V NMR Spectroscopy

^{51}V NMR can be applied to intrinsically diamagnetic oxidation states, that is V(V) 3d^0, V(-I) 3d^6 and V(-III) 3d^8, and to oxidation states which can be diamagnetic under certain conditions: low spin V(I) 3d^4, low spin V(III) 3d^2 and dinuclear anti-ferromagnetically coupled V(IV) 3d^1. NMR is almost exclusively applied to ^{51}V, because, as seen before, this is the most abundant isotope and because of the sufficiently more favorable properties of this nucleus (high receptivity relative to proton and rather small quadrupole moment). ^{51}V is a quadrupolar nucleus, with short relaxation times, which gives rather large linewidth. This disadvantage is balanced by the broad shift, which covers, for inorganic vanadates, the range from +2600 to -900 ppm [5].

Depending on the V concentration, pH and ionic strength, aqueous solutions obtained dissolving a V(V) salt can contain a great number of vanadate species in different protonation and condensation states.

Vanadate, VO_4^{3-}, exists only in very basic solutions, while around neutral pH exist $H_2VO_4^-$ and HVO_4^{2-} (pK_a $H_2VO_4^-$ = 7.8). These two species tend to form dimeric ($H_2V_2O_7^{2-}$, $HV_2O_7^{3-}$), tetrameric ($V_4O_{12}^{4-}$), pentameric ($V_5O_{15}^{5-}$) and decameric structures ($HV_{10}O_{28}^{5-}$, $H_2V_{10}O_{28}^{4-}$, $H_3V_{10}O_{28}^{3-}$); all these species were characterized for their chemical shift, pH dependence and pK_a values. In these species VO_4 tetrahedra share corners except in decavanadate where the coordination geometry of V is octahedral. In acidic solutions it is believed that H_3VO_4 does not exist while $[VO_2(H_2O)_4]^+$ (= $H_4VO_4^+ \cdot 2H_2O$) is more likely [6].

Therefore the speciation depends on the V concentration (and ionic strength) and at millimolar concentration at physiological pH monomeric vanadates barely reach 50% of total vanadium and coexist with dimeric and tetrameric species. Only at micromolar or lower concentrations monomeric vanadates are solely present in aqueous solution.

Figure 1 depicts the speciation of aqueous V(V) solutions, at pH 7.4, as a function of the total V concentration.

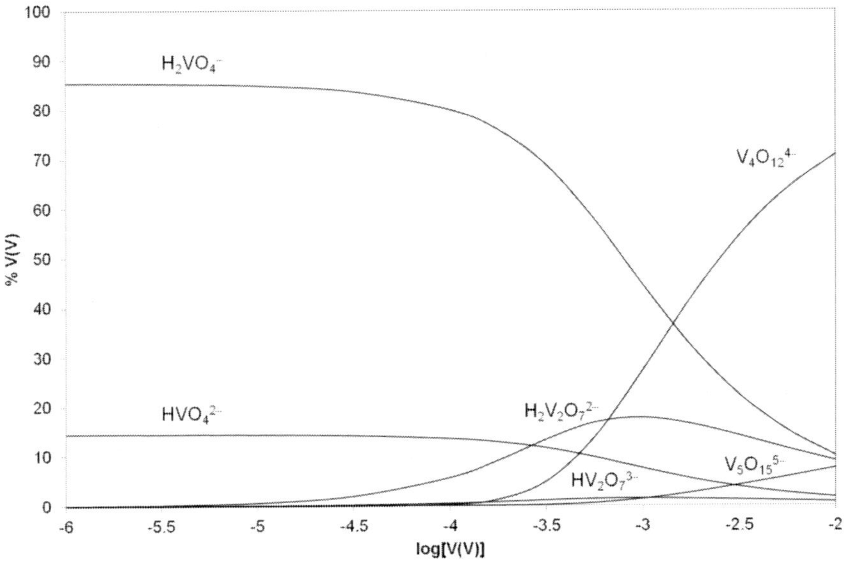

Figure 1. Distribution diagram of vanadate(V) species as a function of total vanadium concentration at pH 7.4. The species distribution was calculated using the stability constants reported in ref. [7].

The detailed characterization of aqueous vanadate solutions was possible because of a very powerful spectroscopic technique, that is, ^{51}V NMR. The reference standard for ^{51}V NMR chemical shifts is $VOCl_3$, for which the chemical shift is set to zero. An alternative standard is a 1 M solution of vanadate at pH 12, which contains anions VO_4^{3-} (= -535.7 ppm) and $V_2O_7^{4-}$(= -559.0 ppm). When the molecules or ions containing the ^{51}V nucleus can freely rotate in solution, the shielding σ has mainly two contributions $\sigma = \sigma_{dia} + \sigma_{para}$ (there is also another term σ_{nl}, where nl stands for non-local contributions, but its value is usually negligibly small in ^{51}V NMR). The diamagnetic contribution is dominated by the core electrons; σ_{dia} is an essentially *constant* term, and thus hardly contributes to any variations in shielding. σ_{para} refers to electronically excited 'paramagnetic' states, arising from an electronic status where an electron from a doubly occupied orbital (diamagnetic S = 0 ground state) is excited to an originally unoccupied orbital. The term σ_{para} thus depends on: i) the strength of the ligand field, ii) the nephelauxetic effect, or softness, or polarisability of the ligands, and iii) the covalency of the metal ligand bond. σ_{para} provides a negative contribution to shielding, that is, σ_{para} is a deshielding term. Thus, as σ_{para} becomes small, overall shielding increases. A small σ_{para} term, and hence effective overall shielding, is to be expected as: i) the ligand field is strong, ii) the ligands are soft, and iii) the vanadium-to-ligand bonds have a high covalent character.

For inorganic vanadium compounds at the high-field and at low-field limits there are vanadium compounds with particularly soft (low electronegativity) and particularly hard (high electronegativity) ligands, respectively. There are a number of empirical rules which can be cited for the dependence of the shielding [8] from electronegativity, coordination number and geometry, steric effects, etc. The chemical shifts and their pH-dependence of inorganic vanadates (see above) can be easily explained following these experimental rules [8].

Linewidths are influenced by a series of variables and therefore is difficult to give linewidth ranges for certain types of complexes. In fact linewidths are influenced by the viscosity of the solvent (which in turn depends on the temperature and the concentration of the V compounds), ligand bulkiness, symmetry of the complexes (O_h, T_d, etc.), ring strains in chelated complexes, hard or soft character of the donor atoms.

EPR Spectroscopy

The information attainable from an EPR spectrum of a VO^{2+} complex depends on if one deals with a liquid solution or a frozen solution containing the dissolved complex. In a liquid solution the fast tumbling of the complexes averages the parameters and, therefore, isotropic parameters are obtained and g_{iso} and A_{iso} can be measured from the spectrum. In a frozen solution anisotropic parameters are obtained and, as a general case, three *g* (g_x, g_y and g_z) and *A* (A_x, A_y and A_z) values can be measured.

With a rhombic EPR symmetry (three *g* and *A* values), the associated geometry must belong to one of the point groups: D_{2h}, C_{2v}, or D_2. A structure which belongs to a more symmetric arrangement, e.g., D_{4h}, is strictly axial, and only two *g* and *A* values are expected ($A_z = A_{||}$; $A_x = A_y = A_{\perp}$) [9].

Wuthrich [10] showed that for some complexes, the isotropic coupling constant, A_0 or A_{iso}, depends on the ligands coordinated in the equatorial plane of the VO^{2+} ion (L_1–L_4 donors in Scheme 2).

Subsequently, Chasteen [11] introduced the "additivity relationship", an empirical rule which assigns a contribution to the ^{51}V hyperfine splitting constant (hsc) to each donor group coordinated in the equatorial plane of a VO^{2+} complex. With this relationship both A_{iso} and A_z, the anisotropic hsc along the z axis measured for a polycrystalline sample or on a frozen solution, can be calculated once the contributions of the donor groups are known.

The main limitation of this relationship is that the comparison should be made only for complexes which have the same geometry, because significant distortion can effectively affect the results of the comparison. Moreover the structure of the complex used to get the contribution of a certain donor group should be known, and this information is not frequently available. For example let's suppose to use the complex $[VO(ox)_2(H_2O)]^{2-}$ (ox = oxalate) to get the contribution of a carboxylate group, and assume that the crystal structure of this complex is not available. If we suppose that four carboxylate groups are coordinated in the equatorial plane of the complex, we certainly obtain a wrong contribution for the carboxylate group because in the real structure a water molecule occupies the fourth equatorial position and one of the carboxylates is coordinated in the axial position (see below).

A significant step in the assignment of the coordination modes to VO^{2+} complexes based on EPR spectroscopy was the comparison of the experimental A_{iso} and A_z values with those calculated with DFT methods. In fact hsc values can now be obtained with a certain degree of reliability with DFT calculations. Garribba and coworkers [12], [13], [14] performed a great number of calculations in order to establish the basis set and functional which give the best agreement between the experimental and calculated hsc values of VO^{2+} complexes. Therefore nowadays, after performing DFT calculations and comparing the obtained hsc values with the experimental ones, it is relatively easier to hypothesize the coordination mode of an unknown VO^{2+} complex.

EPR spectroscopy can be used for the characterization of other transition metal ions, for example Cu(II). Copper(II) has a $3d^9$ electronic configuration and thus, having only one unpaired electron, can be easily studied with this spectroscopic technique. The isotopic distribution of this metal ion is not as favorable as in the case of V, since it has two isotopes, ^{63}Cu and ^{65}Cu both with nuclear spin I = 3/2, which represent 69.2 and 30.8%, respectively. When the isotopic distribution is not favorable, as in the case of copper, or when the most abundant isotope does have a nuclear spin equal to zero, as in the case of chromium (^{50}Cr 4.3%, ^{52}Cr 83.8%, ^{54}Cr 2.4% all with I = 0; ^{53}Cr 9.5% with I = 3/2), the use of isotopically enriched samples is strongly suggested. In the case of Cu(II) an empirical rule analogous to the described above "additivity relationship" has not been developed. In fact even if the g_z or g_{iso} values measured from EPR spectra are indicative of the coordination environment of the metal ion, the A_z or A_{iso} values are both sensitive to the type of coordinated donor atoms but also to the distortion of the complexes. The coordination geometry of copper(II) can distort very easily and even small distortions give rise to significant changes of A values. A parameter that can be used to compare the distortion degree of Cu(II) complexes with the same coordination mode is the g_z/A_z ratio.

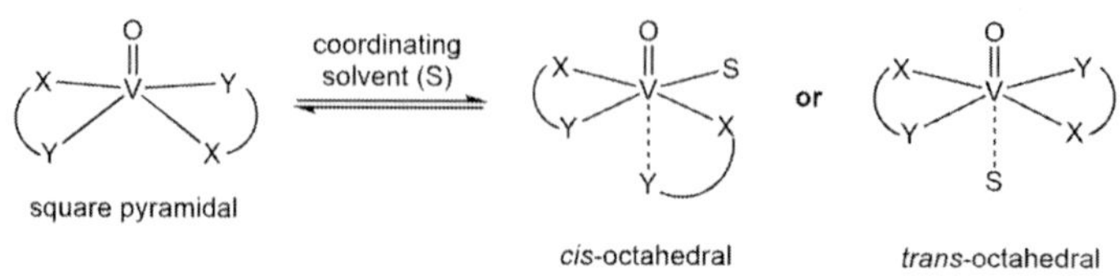

Scheme 4. Equilibria between different coordination isomers for bis-chelated oxidovanadium(IV) complexes.

Most of the bis-chelated oxidovanadium(IV) complexes with bidentate ligands are pentacoordinated with, more or less distorted, square pyramidal geometry. Examples are $VO(acac)_2$ (acac = acetylacetonate) [15], $VO(ma)_2$ (ma = maltolate) [16], $VO(cat)_2$ (cat = catecholate) [17], $VO(8\text{-hq})_2$ (8-hydroxyquinolinate) [18].

There are a few examples of structures having a aqua ligand as the sixth donor coordinated in the equatorial plane as in the case of $[VO(ox)_2(H_2O)]^{2-}$ [19], where ox is oxalate. The structures in solution were supposed to be square pyramidal except in some cases where it was demonstrated that the coordination geometry was *cis*-octahedral (Scheme 4). For example in the case of the complex with maltolate, the structure in the solid state is square pyramidal, while in aqueous solution only the *cis*-octahedral complex is present. In the case of $VO(dhp)_2$, where dhp is 1,2-dimethyl-3-hydroxy-4(1*H*)-pyridinone, there is in aqueous solution an equilibrium between square pyramidal and *cis*-octahedral complexes (Scheme 4) [20].

It was demonstrated that bidentate ligands with higher basicity favor the square pyramidal structures, while less basic donors stabilize the *cis*-octahedral geometry [21]. More recently it was demonstrated that with 2-hydroxypyridine-N-oxide and 2-mercaptopyridine-N-oxide an equilibrium between these two forms exists and this is influenced by the solvent and temperature. Coordinating solvents (H_2O, DMSO, DMF) and lower temperature favor the *cis*-octahedral geometry, while non-coordinating solvents (toluene, CH_2Cl_2, $CHCl_3$) and higher temperature favor the square planar coordination.

The variation in coordination geometry is accompanied by a color change and therefore this reaction can be considered a didactic example to illustrate equilibria between different coordination isomers in transition metal complexes [22]. Other examples are well known in the literature as the case of the equilibrium between tetrahedral and octahedral Co(II) complexes where the variation in the coordination geometry is accompanied by a color change from blue to pink [23]. Another example is given by octahedral and square planar Ni(II)-trien (trien = triethylenetetramine) complexes; in this case the equilibrium can be shifted towards the tetracoordinated square planar species with an increase of the ionic strength of the solution [24]. All these examples show that equilibria involving different species can exist in aqueous solution, while in the solid state only one of the species can be isolated and characterized, and not necessarily this is the most abundant species in solution.

UV-Vis Spectroscopy

As previously stated, most of the bis-chelated oxidovanadium(IV) complexes exist as, more or less distorted, square pyramidal structures. The distortion degree depends on the donor atoms of the bidentate ligand. When the two donor atoms of the bidentate ligand are different, and also the two donor atom-metal bonds are different, the structure becomes distorted towards the trigonal bipyramid. The distortion degree can be quantified and expressed as a difference between two angles (Scheme 5) [25]. The distortion of a square pyramidal structure towards the trigonal bipyramid is accompanied by a loss of degeneracy of the d_{xz}, d_{yz} orbitals (Scheme 2) and therefore four absorption bands are expected in the UV-Vis spectra of these complexes. For VO^{2+} complexes of α-hydroxyacids the distortion degree was put in relationship with the steric hindrance of the R group on the C atom bearing the alcohol function [26]. The loss of orbital degeneracy is usually accompanied by an increase in the number of bands in the UV-Vis spectrum and therefore is easily detected with UV-Vis absorption spectrophotometry.

An empirical rule analogous to the "additivity relationship," described above when discussing EPR spectroscopy of V complexes, was developed for the assignment of the coordination modes of Cu(II)-peptide complexes based on UV-Vis spectroscopy. Cu(II) is able to induce the deprotonation of amide nitrogens of peptides; an analogous behavior is shown by other metal ions with high affinity for nitrogen donors, for example Ni(II) and Pd(II). This empirical rule, originally developed by Billo [27] and then modified by Sigel and Martin [28], assigns to each donor of a peptide molecule, coordinated to the Cu(II) ion, a different contribution and therefore allows for obtaining the wavelength of the absorption maximum of the UV-Vis spectrum.

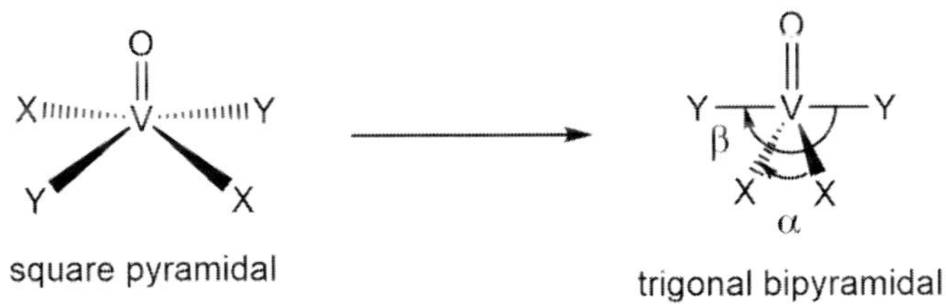

Scheme 5. Distortion of a square pyramidal geometry ($\tau = 0$) towards a trigonal bipyramidal one ($\tau = 1$). The distortion degree can be measured by $\tau = (\beta - \alpha)/60$.

pH-Potentiometric Data

pH-potentiometric titrations provide the composition and protonation degree of each species detected in solution; therefore, pH-potentiometric titrations are usually used in combination with spectroscopic methods to assign coordination modes to the species in solution. The main limitation of pH-potentiometric titration is that the formation of a certain species must be accompanied by a proton loss from the ligand. When the complexes are formed with already deprotonated ligands there is no measurable effect in the titration and therefore these species are not detected. The information given by this technique is limited to the composition and protonation degree. For example, if we have in solution a species with $VOLH_{-1}$ composition and proton is not lost from the ligand (where H_{-1} indicated the proton that was not lost when titrating the ligand alone, so these can be referred to the protons lost from the ligand only in the presence of the metal ion[1], or to the protons lost from coordinated water molecules), it is likely that the species in solution is VOL(OH) where the proton is lost from a coordinated water molecule, but it is more

[1] Amide nitrogen groups –(C=O)–NH– do not undergo protonation-deprotonation processes in aqueous solution in the measurable pH range (0–14), but can be deprotonated in the presence of metal ions which have a high affinity for nitrogen donors, like Cu(II), Ni(II), Pd(II), Pt(II), etc. In these cases the ligand with deprotonated amide nitrogen is indicated with LH_{-1}, because this proton is not lost in the presence of the ligand alone.

difficult to establish if the real complex is VOL(OH), $(VO)_2L_2(OH)_2$, … or $(VO)_nL_n(OH)_n$. Titrations at different concentrations should be performed in order to have a deeper knowledge of such kind of systems, but it is difficult to perform pH-potentiometric measurements in very diluted solutions. EPR spectroscopy cannot be applied to the detection of polynuclear complexes because these are usually EPR-silent. In analogy with the chemistry of vanadium(V) in aqueous solutions, it is likely that di-, tri-, tetra-, penta-, deca-nuclear hydrolytic species also exist in the aqueous chemistry of oxidovanadium(IV). However, in this case the lack of a specific technique allowing for detecting such species makes the aqueous chemistry of this latter less characterized (Figure 2) in comparison with that of V(V) for which ^{51}V NMR is very useful in this regard. Therefore, extending these considerations to other transition metal ions, it is desirable always to use pH-potentiometric data in combination with other spectroscopic methods in order to have a more detailed description of the systems.

Transition metal ions exist in aqueous solutions as aquaions of the general formula $[M(H_2O)_n]^{z+}$, where n usually 6 or 4 and z is the charge of the transition metal ion.

These complexes behave as Brønsted acids, that is, show acidic features releasing a proton: $[M(H_2O)_n]^{z+}$ λ $[M(H_2O)_{n-1}(OH)]^{(z-1)+} + H^+$.

The strength of aquaions as acids increases with increasing the positive charge of the metal ions and with decreasing its ionic radius. For example pK_a of $[Fe(H_2O)_6]^{3+}$ is 2.19 [33].

For the less electronegative metals (with $\chi < 1.5$ on Pauling scale) there is a good correlation between the pK_a and the ratio z^2/r (where z is the charge and r the radius of the metal). For more electronegative metals, another empirical relationship must be used to take into account that the metal–OH_2 bond cannot be described exclusively as ionic, but has some character of covalency.

Transition metal aquaions exhibit acidic features and therefore their aqueous solutions have acidic pH values; if the pH of these solutions is increased OH^- ions will replace the water molecules in the coordination sphere of $[M(H_2O)_n]^{z+}$, therefore a series of complexes will form:

$[M(H_2O)_{n-1}(OH)]^{(z-1)+}$, $[M(H_2O)_{n-2}(OH)_2]^{(z-2)+}$, $[M(H_2O)_{n-3}(OH)_3]^{(z-3)+}$, etc. depending on the charge z. When the number of OH^- ions matches the charge z, the hydroxide is neutral and often poorly soluble forming colloidal precipitates.

The equilibria involving precipitates are heterogeneous since they involve two different phases: the liquid solution and the solid precipitate. Heterogeneous equilibria are not governed by the same laws which control homogeneous ones since the exchange between the two phases only takes place at the interface.

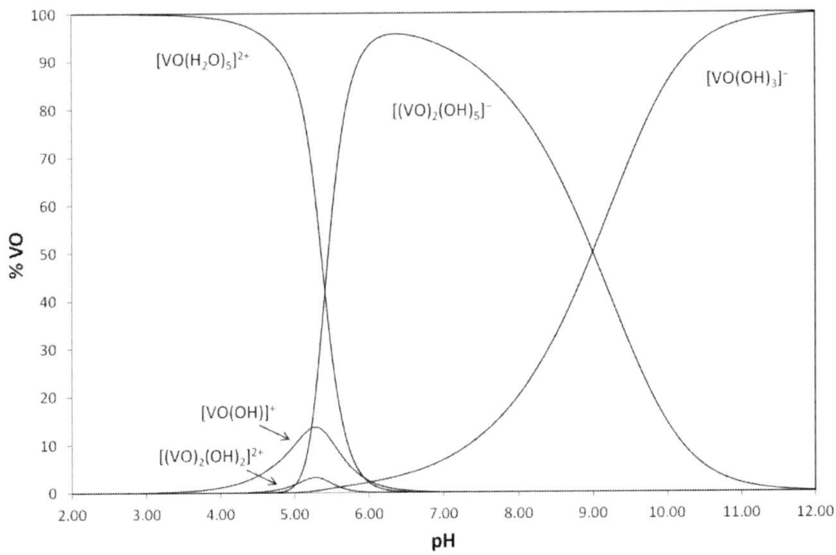

Figure 2. Distribution diagram of the species of aqueous VO^{2+} solution at 10 μM. The following hydroxide complexes were assumed: $[VO(OH)]^+$ (log $\beta_{10-1} = -5.94$), $[(VO)_2(OH)_2]^{2+}$ (log $\beta_{20-2} = -6.95$), with stability constants calculated from the data of ref. [29], $[VO(OH)_3]^-$ (log $\beta_{10-3} = -18.0$) and $[(VO)_2(OH)_5]^-$ (log $\beta_{20-5} = -22.0$) with those of refs. [30] and [31]. Besides these species $VO(OH)_2$ exists with $K_{sp} = 6.6\text{x}10^{-23}$ [32].

Depending on the specific metal ion these hydroxides can dissolve or not in an excess of base, that is, can show or not amphoteric character. When this is the case these metals form negatively charged hydroxide-complexes.

Non-Innocent Ligands

Jørgensen pointed out in 1966 that "ligands are innocent when they allow oxidation states of the central atoms to be defined", a definition which puts in evidence that non-innocent behavior depends on the metal as well as the ligand [34]. Often the term non-innocent is used to indicate ligands that in a complex are redox-active; therefore the term non-innocent is often and improperly used to indicate dioxolenes. The term dioxolene is used to indicate the redox series catecholate (the dianion of benzene-1,2-diol), semiquinonate anion and 1,2-benzoquinone (Scheme 6).

In complexes where the redox potentials of the ligands and those of the metals are well separated, and therefore the redox processes are unambiguously assigned to the metals or to the ligands, the oxidation state of the central atoms can be defined. In these cases the ligands should be defined as innocent ones.

When catechol derivatives form complexes with metal ions which can undergo reduction processes, the assignment of the oxidation state to the metal ion is not banal. For example if we have V(V), Cu(II) or Fe(III) as metal ions, these can in principle be reduced to V(IV), Cu(I) and to Fe(II); if the reduction potentials of V(V), Cu(II) and Fe(III) ions are similar to the oxidation potentials of the catechol derivatives then the assignment of the oxidation state to the metal ions is ambiguous. In fact depending on the conditions V(V), Cu(II) and Fe(III) can maintain their original oxidation state (the same happens in this case for the ligands), or can undergo reduction processes with the ligands with intramolecular exchange of electrons.

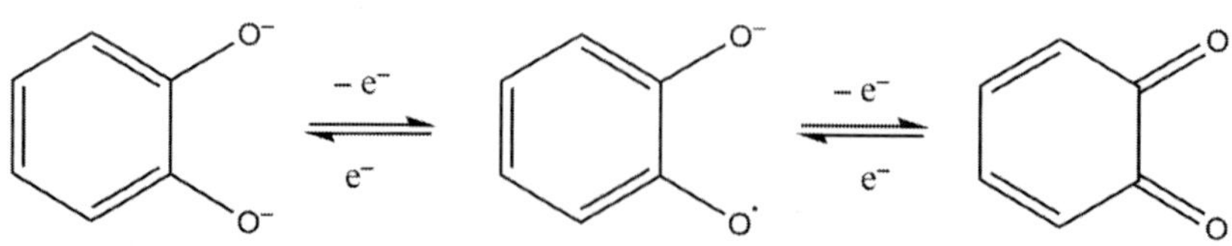

Scheme 6. Oxidation of catecholate dianion to semiquinonate anion and to 1,2-benzoquinone.

Depending on the substituents on the aromatic ring, the oxidation potential of catechol derivatives can change and become comparable to the reduction potential of V(V), Cu(II) or Fe(III) (which also depends on the solvent and the other ligands eventually coordinated to the metal ions), making possible the intramolecular exchange of electrons between the ligand and the metal.

CONCLUSION

The examples given in this chapter demonstrate that solution chemistry of transition metal ions can be very complicated because different equilibria can take place depending on the metal ion, its thermodynamic stability, its kinetic lability, the coordinated ligands, the solvents, etc. These examples also show that even if a specific technique is available to study a particular metal ion, this certainly has limitations and therefore the results of its application can give a partial and incomplete picture of the real situation. The combined application of at least two complementary and independent techniques is certainly the best option to study the solution chemistry of transition metal ions. In this way the system under investigation is observed by, at least, two different points of view giving more reliable results.

The main limitation of this approach is that for certain metal ions very few techniques are available and then, in these cases, it is difficult to give a clear and complete description of the speciation in solution.

REFERENCES

[1] M. Nic, J. Jirat, B. Kosata, updates compiled by A. Jenkins, IUPAC. *Compendium of Chemical Terminology*, 2nd ed. (the "Gold Book"), 2006.

[2] H. Bethe, Term splitting in crystals, *Annals of Physics*, 395 (1929) 133-208.

[3] R. G. Pearson, Hard and Soft Acids and Bases, *Journal of the American Chemical Society*, 85 (1963) 3533-3539.

[4] S. F. Lincoln, D. T. Richens, A. G. Sykes, 1.25 - Metal Aqua Ions A2 - McCleverty, Jon A, in: T. J. Meyer (Ed.) *Comprehensive Coordination Chemistry II,* Pergamon, Oxford, 2003, pp. 515-555.

[5] D. Rehder, T. Polenova, M. Bühl, Vanadium-51 NMR, in: G. A. Webb (Ed.) *Annual Reports on NMR Spectroscopy*, Academic Press 2007, pp. 49-114.

[6] D. C. Crans, J. J. Smee, E. Gaidamauskas, L. Yang, The Chemistry and Biochemistry of Vanadium and the Biological Activities Exerted by Vanadium Compounds, *Chemical Reviews*, 104 (2004) 849-902.

[7] H. Schmidt, I. Andersson, D. Rehder, L. Pettersson, A Potentiometric and ^{51}V NMR Study of the Aqueous $H^+/H_2VO_4^-/H_2O_2$/L-α-Alanyl-L-histidine System, *Chemistry – A European Journal*, 7 (2001) 251-257.

[8] D. Rehder, Physico-Chemical Methods for the Characterisation of Native and Model Vanadium Compounds, *Bioinorganic Vanadium Chemistry,* John Wiley & Sons Ltd, 2008, pp. 53-86.

[9] C. D. Garner, D. Collison, F. E. Mabbs, Methods for the Spectroscopic Characterization of Vanadium Centers in Biological and Related Chemical Systems, in: H. Sigel, A. Sigel (Eds.) *Metal Ions in Biological Systems*, Marcel Dekker, New York, 1995, pp. 617-670.

[10] K. Wüthrich, Electron spin resonance studies of VO2 + complex compounds in aqueous solution II, *Helvetica Chimica Acta*,48 (1965) 1012-1017.

[11] N. D. Chasteen, Vanadyl(IV) EPR Spin Probes Inorganic and Biochemical Aspects, in: L. J. Berliner, J. Reuben (Eds.) *Biological Magnetic Resonance: Volume 3*, Springer US, Boston, MA, 1981, pp. 53-119.

[12] G. Micera, E. Garribba, The effect of the functional, basis set, and solvent in the simulation of the geometry and spectroscopic properties of $V^{IV}O^{2+}$ complexes. chemical and biological

applications, *International Journal of Quantum Chemistry*, 112 (2012) 2486-2498.

[13] S. Gorelsky, G. Micera, E. Garribba, The Equilibrium Between the Octahedral and Square Pyramidal Form and the Influence of an Axial Ligand on the Molecular Properties of $V^{IV}O$ Complexes: A Spectroscopic and DFT Study, *Chemistry – A European Journal*, 16 (2010) 8167-8180.

[14] G. Micera, E. Garribba, On the prediction of ^{51}V hyperfine coupling constants in $V^{IV}O$ complexes through DFT methods, *Dalton Transactions*, (2009) 1914-1918.

[15] R. P. Dodge, D. H. Templeton, A. Zalkin, Crystal Structure of Vanadyl Bisacetylacetonate. Geometry of Vanadium in Fivefold Coordination, *The Journal of Chemical Physics*, 35 (1961) 55-67.

[16] C. Orvig, P. Caravan, L. Gelmini, N. Glover, F. G. Herring, H. Li, J. H. McNeill, S. J. Rettig, I. A. Setyawati, Reaction chemistry of BMOV, bis(maltolato)oxovanadium(IV), a potent insulin mimetic agent, *Journal of the American Chemical Society*, 117 (1995) 12759-12770.

[17] S. R. Cooper, Y. B. Koh, K. N. Raymond, Synthetic, structural, and physical studies of bis(triethylammonium) tris(catecholato) vanadate (IV), potassium bis(catecholato)oxovanadate(IV), and potassium tris (catecholato)vanadate (III), *Journal of the American Chemical Society,* 104 (1982) 5092-5102.

[18] M. Shiro, Q. Fernando, Structures of two five-coordinated metal chelates of 2-methyl-8-quinolinol, *Analytical Chemistry*, 43 (1971) 1222-1230.

[19] R. E. Oughtred, E. S. Raper, H. M. M. Shearer. The crystal structure of diammonium bis(oxalato)monoaquaoxovanadate(IV) monohydrate. *Acta Crystallographica Section B*, 32 (1976) 82-87.

[20] P. Buglyó, T. Kiss, E. Kiss, D. Sanna, E. Garribba, G. Micera, Interaction between the low molecular mass components of blood serum and the VO(IV)-DHP system (DHP = 1,2-dimethyl-3-hydroxy-4(*1H*)-pyridinone), Journal of the Chemical Society, *Dalton Transactions*, (2002) 2275-2282.

[21] D. Sanna, P. Buglyó, L. Bíró, G. Micera, E. Garribba, Coordinating Properties of Pyrone and Pyridinone Derivatives, Tropolone and Catechol toward the VO^{2+} Ion: An Experimental and Computational Approach, *European Journal of Inorganic Chemistry*, 2012 (2012) 1079-1092.

[22] V. Ugone, E. Garribba, G. Micera, D. Sanna, Equilibrium between Different Coordination Geometries in Oxidovanadium(IV) Complexes, *Journal of Chemical Education*, 92 (2015) 1098-1102.

[23] C. E. Ophardt, Cobalt complexes in equilibrium, *Journal of Chemical Education*, 57 (1980) 453.

[24] P. Paoletti, L. Fabbrizzi, R. Barbucci, Thermochemistry of metal-polyamine complexes, *Inorganica Chimica Acta Reviews*, 7 (1973) 43-68.

[25] C. R. Cornman, K. M. Geiser-Bush, S. P. Rowley, P. D. Boyle, Structural and Electron Paramagnetic Resonance Studies of the Square Pyramidal to Trigonal Bipyramidal Distortion of Vanadyl Complexes Containing Sterically Crowded Schiff Base Ligands, *Inorganic Chemistry,* 36 (1997) 6401-6408.

[26] E. Garribba, G. Micera, A. Panzanelli, D. Sanna, Electronic Structure of Oxovanadium(IV) Complexes of α-Hydroxycarboxylic Acids, *Inorganic Chemistry*, 42 (2003) 3981-3987.

[27] E. J. Billo, Copper(II) chromophores and the rule of average environment, *Inorganic and Nuclear Chemistry Letters*, 10 (1974) 613-617.

[28] H. Sigel, R. B. Martin, Coordinating properties of the amide bond. Stability and structure of metal ion complexes of peptides and related ligands, *Chemical Reviews,* 82 (1982) 385-426.

[29] R. P. Henry, P. C. H. Mitchell, J. E. Prue, Hydrolysis of the oxovanadium(IV) ion and the stability of its complexes with the 1,2-dihydroxybenzenato(2-) ion, *Journal of the Chemical Society, Dalton Transactions,* (1973) 1156-1159.

[30] A. Komura, M. Hayashi, H. Imanaga, Hydrolytic Behavior of Oxovanadium(IV) Ions, *Bulletin of the Chemical Society of Japan*, 50 (1977) 2927-2931.

[31] L. F. Vilas Boas, J. Costa Pessoa, Vanadium, in: G. Wilkinson, R. D. Gillard, J. A. McCleverty (Eds.) *Comprehensive Coordination Chemistry*, Pergamon Press, Oxford, 1987, pp. 453-583.

[32] K. Akitoshi, H. Mutsuo, I. Hiroto, Hydrolytic Behavior of Oxovanadium(IV) Ions, *Bulletin of the Chemical Society of Japan*, 50 (1977) 2927-2931.

[33] D. W. Barnum, Hydrolysis of cations. Formation constants and standard free energies of formation of hydroxy complexes, *Inorganic Chemistry*, 22 (1983) 2297-2305.

[34] C. K. Jørgensen, Differences between the four halide ligands, and discussion remarks on trigonal-bipyramidal complexes, on oxidation states, and on diagonal elements of one-electron energy, *Coordination Chemistry Reviews*, 1 (1966) 164-178.

In: Solution Chemistry
Editor: Yongliang Xiong
ISBN: 978-1-53613-101-7

Chapter 8

MODELING ACTINIDE SOLUBILITIES IN ALKALINE TO HYPERALKALINE SOLUTIONS: SOLUBILITY OF $Am(OH)_3(s)$ IN KOH SOLUTIONS

Yongliang Xiong*
Sandia National Laboratories (SNL),
Defense Waste Management Programs, Carlsbad, NM, US

ABSTRACT

In this work, a Pitzer model is developed for the $K^+(Na^+)$—$Am(OH)_4^-$—Cl^-—OH^- system based on $Am(OH)_3(s)$ solubility data in highly alkaline KOH solutions. Under highly alkaline conditions, the solubility reaction of $Am(OH)_3(s)$ is expressed as:

$$Am(OH)_3(s) + OH^- = Am(OH)_4^-$$

Solubilities of $Am(OH)_3(s)$ based on the above reaction are modeled as a function of KOH concentrations. The stability constant for $Am(OH)_4^-$

* Corresponding Author Email: yxiong@sandia.gov.

is evaluated using $Am(OH)_3(s)$ solubility data in KOH solutions up to 12 $mol \cdot kg^{-1}$ taken from the literature. The Pitzer interaction parameters related to $Al(OH)_4^-$ are used as analogs for the interaction parameters involving $Am(OH)_4^-$ to obtain the stability constant for $Am(OH)_4^-$. The $\log_{10} K^0$ for the reaction is -11.34 ± 0.15 (2σ).

Keywords: nuclear waste management, hydrolysis, Pitzer model, salt formations, high level tank waste, Hanford site, Savannah River site, Thermodynamic modeling

INTRODUCTION

Knowledge of actinide speciation and solubilities in highly alkaline solutions is important to nuclear waste management. For example, tank wastes such as those at the Savannah River site and the Hanford site, contain solutions with high concentrations of bases (e.g., Delegard, 1987; Karraker, 1995; Rudisill et al., 2004, 2010; King et al., 2010; Rudisill and Thompson, 2011; Delegard and Jones, 2015; King and Hobbs, 2016). Also, solutions in equilibrium with cements in geological repositories would be highly alkaline with concentrations of OH^- on the order of hundreds of milli-molars in some repository designs (Vitorge and Tran The, 1991). For instance, in the supercontainer design for the geological disposal of spent nuclear fuel in Belgium, waste is inserted into a canister surrounded by a carbon steel overpack, a buffer and then stainless steel envelope (Cachoir et al., 2016). Ordinary Portland cement based concrete is used as the buffer in the Belgian repository design (Loida et al., 2009; Mennecart et al., 2012; Cachoir et al., 2014, 2016). Concrete barriers are also employed in the French repository design (e.g., De Windt et al., 2004; Trotignon et al., 2006). The concrete buffer is expected to produce a high pH environment for the long-term.

Predictions of actinide solubilities in these highly alkaline solutions require a thermodynamic model applicable to high ionic strengths with high base concentrations. Current speciation schemes for actinides do not extend into the highly alkaline region (Silva et al., 1995). If modeling actinide solubility in the environments mentioned above is required, a parameterized speciation scheme for this region needs to be developed.

The objective of this work is to develop a Pitzer model for the $K^+(Na^+)$—$Am(OH)_4^-$—Cl^-—OH^- system based on $Am(OH)_3(s)$ solubility data in highly alkaline solutions. The model is applicable to hyperalkaline solutions.

METHODS

The platform for modeling in this chapter is the computer program EQ3/6 Version 8.0a (Wolery et al., 2010; Xiong, 2011a). The EQ3/6 code has been successfully used as a modeling platform in a number of previous studies including those at ambient temperature (e.g., De Windt et al., 2004; Xiong et al., 2013a, 2013b; Xiong, 2015) and those at elevated temperatures up to 523.15 K (Xiong, 2013a, 2013b, 2014). The database used for the modeling is summarized in Xiong et al., (2010).

In the modeling, experimental data selected are used to generate EQ3/6 input files in the first step. In the second step, the values of the targeted parameter (e.g., $\log K_1^0$) are changed until the squared difference between experimental values (e.g., $m_{\Sigma Am}$) and model-produced values predicted by the computer code is minimized. In the third step, the final selected value for the targeted parameter is verified to see how it reproduces experimental data.

MODELING RESULTS AND POTENTIAL APPLICATIONS

Modeling Results

It has been observed that $Am(OH)_4^-$ is the dominant species in highly alkaline solutions (Vitorge and Tran The, 1991). Under highly alkaline conditions, the solubility reaction for $Am(OH)_3(s)$ is expressed as,

$$Am(OH)_3(s) + OH^- \rightleftarrows Am(OH)_4^- \quad (1)$$

Vitorge and Tran The (1991) conducted solubility experiments at 25°C on $Am(OH)_3(s)$ in KOH solutions up to 10 mol•dm^{-3}. Here, the solubility data from Vitorge and Tran The (1991) are modeled to obtain an equilibrium constant for Reaction (1). In the modeling, the experimental data of Vitorge and Tran The (1991) on a molar scale are converted to those on a molal scale, based on the density equation from Sőhnel and Novotný (1985).

A Pitzer model is adopted for the modeling and the Pitzer interaction parameters related to $Al(OH)_4^-$ (Wesolowski, 1992; Xiong, 2014) are used as analogs for the interaction parameters involving $Am(OH)_4^-$ to obtain the stability constant for $Am(OH)_4^-$ (see Table 1). These Pitzer parameters have been tested in a high ionic strength range up to 12 mol•kg^{-1} (Wesolowski, 1992; Xiong, 2013b, 2014).

Based on the modeling performed in this work, the equilibrium constant for Reaction (1) is -11.34 ± 0.15 (2σ) in logarithmic units (Table 2). To facilitate its use in various databases, the equilibrium constant for Reaction (1) is converted to other equilibrium constants expressed in Reactions (4) and (5), in accordance with the basis reactions in Reactions (2) and (3) from the DATA0.FM1 (Xiong, 2011b),

$$Am(OH)_3(s) + 3H^+ \rightleftarrows Am^{3+} + 3H_2O(l) \quad (2)$$

$$H_2O(l) \rightleftarrows H^+ + OH^- \quad (3)$$

$$Am(OH)_3(s) + H_2O(l) \rightleftarrows H^+ + Am(OH)_4^- \quad (4)$$

$$Am(OH)_4^- + 4H^+ \rightleftarrows Am^{3+} + 4H_2O(l) \quad (5)$$

The equilibrium constants for the above reactions are also listed in Table 2.

Table 1. The Pitzer interaction parameters at 25°C and 1 bar for the K^+—$Am(OH)_4^-$—Cl^-—OH^- system

Pitzer binary interaction parameters					
Species, *i*	Species, *j*	$\beta^{(0)}$	$\beta^{(1)}$	C^{ϕ}	Reference
K^+	$Am(OH)_4^-$	0.051	0.25	-0.00090	This work [A]
Pitzer mixing interaction parameters (theta and psi parameters)					
Species, *i*	Species, *j*	Species, *k*	θ_{ij}	Ψ_{ijk}	Reference
OH^-	$Am(OH)_4^-$	K^+	0.014	-0.0048	This work[B]
Cl^-	$Am(OH)_4^-$	$K^+(Na^+)$		-0.004857	This work[C]

[A] Using the parameters for Na^+—$Al(OH)_4^-$ from Wesolowski (1992) as analogs.

[B] Using the parameters for OH^-—$Al(OH)_4^-$ and OH^-—$Al(OH)_4^-$— Na^+ from Wesolowski (1992) as analogs.

[C] Using the parameter for OH^-—$Al(OH)_4^-$— Na^+ from Xiong (2014) as an analog.

Table 2. Equilibrium constants at infinite dilution and at 25°C and 1 bar for the K^+—$Am(OH)_4^-$—Cl^-—OH^- system

Reaction	$\log_{10} K^0$	**Reference**
$Am(OH)_3(s) + 3H^+ \rightleftarrows Am^{3+} + 3H_2O(l)$	14.4900	DATA0.FM1 (Xiong, 2011b)
$H_2O(l) \rightleftarrows H^+ + OH^-$	-13.9967	DATA0.FM1 (Xiong, 2011b)
$Am(OH)_3(s) + OH^- \rightleftarrows Am(OH)_4^-$	-11.34 ± 0.15 (2σ)	This work
$Am(OH)_3(s) + H_2O(l) \rightleftarrows H^+ + Am(OH)_4^-$	-25.34 ± 0.15 (2σ)	This work
$Am(OH)_4^- + 4H^+ \rightleftarrows Am^{3+} + 4H_2O(l)$	39.83 ± 0.15 (2σ)	This work

Shown in Figure 1 is the comparison of solubilities of $Am(OH)_3(s)$ predicted by the model developed in this study with the experimental data, as a function of ionic strength. The experimental data are from Vitorge and Tran The (1991). There are five sets of experimental data in Vitorge and Tran The (1991). The first two sets of experimental data are in KOH solutions and lime water without the redox adjustment at 70 and 105 days, respectively (Figure 1). The third set is in KOH solutions and lime water with initial oxidation and subsequent reduction. The last two sets of experimental data are in KOH solutions and cement water with the fourth being started with initial oxidation and subsequent reduction, and with the fifth being without a redox change. The results from the first two sets of experiments with KOH solutions and lime water [i.e., saturated with $Ca(OH)_2(s)$] are used in the model development, and therefore they are model-dependent. The rest of sets are not used in the model development. They are model-independent. All of the experiments of Vitorage and Tran The (1991) approached the equilibrium from the direction of supersaturation. Figure 1 demonstrates that the model developed in this study can reproduce the model-dependent experimental data with good agreement. It is also in good agreement with the model-independent experimental data.

There was a wide range of estimated values for $Am(OH)_4^-$ in the literature. Regarding Reaction (5), Allard et al., (1978) estimated a value of 31 in logarithmic units. Baes and Mesmer (1976) estimated a value of 37.1 in logarithmic units for $Nd(OH)_4^-$. In comparison with the value of 39.83 ± 0.15 (2σ) for $Am(OH)_4^-$ in logarithmic units obtained in this study (Table 2), the estimated value provided by Baes and Mesmer (1976) is closer to the current value based on experimental data.

Potential Applications

Potential, immediate applications of the results obtained in this study would be to model the solubilities of actinides in the +III oxidation state [i.e., An(III)] in a high level tank waste. Potential candidates would be the high level tank waste at the Hanford and Savannah River sites. As indicated by Table 3, the high-level tank waste at the Hanford site has a high NaOH concentration at 4 $mol{\bullet}dm^{-3}$. In addition, the liquid waste has relatively high concentrations of EDTA and citrate. As EDTA and citrate form strong complexes with actinides in the +III oxidation state, the formation constants of EDTA and citrate with An(III) applicable to high ionic strength solutions from Thakur et al., (2012, 2014) need to be used in such a modeling.

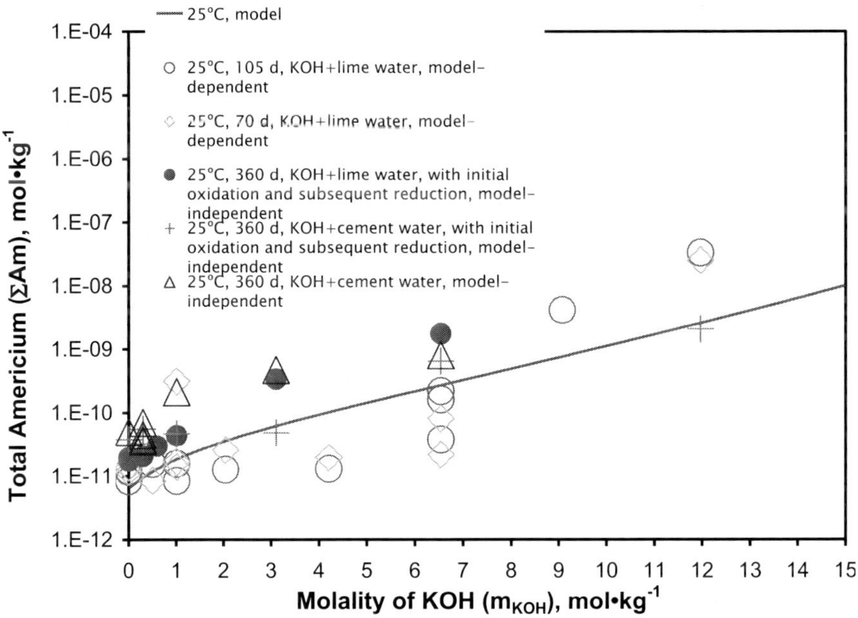

Figure 1. A plot showing the comparison of solubilities of $Am(OH)_3(s)$ predicted by the model developed in this study with the experimental data, as a function of ionic strength. The experimental data are from Vitorge and Tran The (1991).

Table 3. Typical compositions of high-level tank waste at the Hanford site*

Component	Concentration, mol•dm^{-3}
NaOH	4
$NaNO_3$	2
$NaNO_2$	2
$NaAl(OH)_4$	0.5
Na_2CO_3	0.05
Na_2SO_4	0.01
Na_3PO_3	0.01
NaF	0.01
Na_4EDTA	0.05
$Na_3Citrate$	0.03
NaGlycolite	0.1

* From Delegard (1985).

Table 4. Typical compositions of high-level tank waste at the Savannah River site*

Component**	**Concentration, mol•dm^{-3}**
OH^-	0.0001 to 15
NO_3^-	0.1 to 6
NO_2^-	0.1 to 6
$Al(OH)_4^-$	0.001 to 1
CO_3^{2-}	0.001 to 1
SO_4^{2-}	0.001 to 0.5

* From Rudisill et al., (2004).

** The cation is predominantly Na^+.

Similarly, the high-level tank waste at the Savannah River site also has high NaOH concentrations up to 15 mol•dm^{-3} (Table 4). It also has high concentrations of NO_3^-, NO_2^-, and $Al(OH)_4^-$. It also contains relatively high concentrations of CO_3^{2-} and SO_4^{2-}. However, the high-level tank waste at the Savannah River site is different from the high-level tank waste at the Hanford site in that it is poor in organic ligands (cf. Tables 3 and 4).

CONCLUSION

In this study, we have developed a Pitzer model for the $K^+(Na^+)$—$Am(OH)_4^-$—Cl^-—OH^- system based on $Am(OH)_3(s)$ solubility data in highly alkaline solutions. This model is expected to find immediate applications in modeling the solubility of An(III) in high level tank waste. It is anticipated that the model will also be applicable to the modeling of the solubility of An(III) in geological repositories in various geological media including salt formations, where cement is used as one of multiple buffers.

ACKNOWLEDGMENTS

Sandia National Laboratories is a multi-mission laboratory managed and operated by National Technology and Engineering Solutions of Sandia, L.L.C., a wholly owned subsidiary of Honeywell International, Inc., for the U.S. Department of Energy's National Nuclear Security Administration under contract DE-NA-0003525. This research is funded by the Salt R&D programs administered by the Office of Nuclear Energy (NE) of the U.S. Department of Energy. SAND2017-10095B.

REFERENCES

Allard, Bert, Heino Kipatsi, and Börje Torstenfelt, 1978. "Sorption of long half-life radionuclides in clay." *KBS-98, Karnbranslesakerhet,* Stockholm.

Baes, Charles F., and Robert E. Mesmer, 1976. *Hydrolysis of cations.* Wiley.

Cachoir, C., Th Mennecart, and K. Lemmens, 2014. "Effect of cement water on UO_2 solubility." *MRS Online Proceedings Library Archive* 1665: 267-273.

Cachoir, Christelle, Sonia Salah, Thierry Mennecart, and Karel Lemmens, 2016. "Uranium Dissolution in Hyperalkaline TMA-OH Solutions: Preliminary Results." *Procedia Chemistry* 21:306-313.

De Windt, Laurent, Delphine Pellegrini, and Jan Van Der Lee, 2004. "Coupled modeling of cement/claystone interactions and radionuclide migration." *Journal of Contaminant Hydrology* 68:165-182.

Delegard, Calvin H., 1985. *Solubility of $PuO_2{\bullet}xH_2O$ in alkaline Hanford high-level waste solution.* No. RHO-RE-SA-75-P; CONF-850417-31. Rockwell International Corp., Richland, WA (USA). Rockwell Hanford Operations.

Delegard, Calvin H., 1987. "Solubility of $PuO_2{\bullet}xH_2O$ in Alkaline Hanford High-Level Waste Solution." *Radiochimica Acta* 41:11-22.

Delegard, Calvin H., and Susan A. Jones, 2015. *Chemical disposition of plutonium in Hanford site tank wastes*. No. PNNL--23468-Rev. 1. Pacific Northwest National Lab.(PNNL), Richland, WA (United States).

Karraker, D. G., 1995. *Plutonium VI Solubility Studies in Savannah River Site High Level Waste*, WSRC-TR-95-0244, Westinghouse Company.

King, William D., and David T. Hobbs, 2016. *Determining the Release of Radionuclides from Tank 18F Waste Residual Solids: FY2016 Report*. No. SRNL-STI-2016-00432. Savannah River Site (SRS), Aiken, SC (United States). Savannah River National Lab (SRNL).

King, William D., Tommy B. Edwards, David T. Hobbs, and William R. Wilmarth, 2010. "Solubility of Uranium and Plutonium in Alkaline Savannah River Site High Level Waste Solutions." *Separation Science and Technology* 45:1793-1800.

Loida, A., R. Gens, V. Metz, Karel Lemmens, Christelle Cachoir, Thierry Mennecart, and B. Kienzler, 2009. "Corrosion behavior of high burnup spent fuel in highly alkaline solutions." *MRS Online Proceedings Library Archive* 1193.

Mennecart, Th, Christelle Cachoir, and Karel Lemmens, 2012. "UO_2 dissolution in high pH conditions of the Belgian Supercontainer." *MRS Online Proceedings Library Archive* 1475.

Rudisill, T., and M. Thompson, 2011. *Enhanced Chemical Cleaning of SRS Waste Tanks to Improve Actinide Solubility*. No. SRNL-STI-2011-00521. SRS.

Rudisill, Tracys, D. T. Hobbs, T. B. Edwards, 2004. *Preliminary Results from Plutonium/Americium Studies Using Simulated Savannah River Site Waste Solutions*. No. WSRC-TR-2004-00349. Savannah River Site (US).

Rudisill, Tracy S., David T. Hobbs, and Thomas B. Edwards, 2010. "Plutonium solubility in simulated Savannah River Site waste solutions." *Separation Science and Technology* 45:1782-1792.

Silva, R. J., Bidoglio, G., Robouch, P. B., Puigdomènech, I., Wanner, H., and Rand, M. H., 1995. *Chemical thermodynamics of americium* (Vol. 2). NEA OECD, Elsevier, Amsterdam.

Sőhnel, O, Novotný, P., 1985. *Densities of aqueous solutions of inroganic substances*. Elsevier, New York, p. 335.

Thakur, P., Xiong, Y.-L., Borkowski, M., and Choppin, G. R., 2012. "Thermodynamic modeling of trivalent Am, Cm and Eu-citrate complexation in concentrated $NaClO_4$ media." *Radiochimica Acta* 100:1-8.

Thakur, P., Xiong, Y.-L., Borkowski, M., and Choppin, G. R., 2014. "Improved thermodynamic model for interaction of EDTA with trivalent actinides and lanthanide to ionic strength of 6.60 m." *Geochimica et Cosmochimica Acta* 133:299-312.

Trotignon, Laurent, Hugues Peycelon, and Xavier Bourbon, 2006. "Comparison of performance of concrete barriers in a clayey geological medium." *Physics and Chemistry of the Earth, Parts A/B/C* 31: 610-617.

Vitorge, P., Tran The, 1991. *Solubility limits of radionuclides in interstitial water. Americium in cement. Task 3. Characterization of radioactive waste forms. A series of final reports (1985-89)*. No 34. No. EUR-13664. Commission of the European Communities, 1991.

Wesolowski, D. J., 1992. "Aluminum speciation and equilibrium in aqueous solution: I. The solubility of gibbsite in the system Na-K-Cl-OH-$Al(OH)_4$ from 0°C to 100°C." *Geochimica et Cosmochimica Acta* 56:1065-1091.

Wolery, T. W., Xiong, Y.-L., Long, J. J., 2010. *"Verification and Validation Plan/Validation Document for EQ3/6 Version 8.0a for Actinide Chemistry, Document Version 8.10."* ERMS 550239, Carlsbad, New Mexico, Sandia National Laboratories.

Xiong, Y.-L., 2011a. *"Verification and Validation Plan/Validation Document for EQ3/6 Version 8.0a for Actinide Chemistry, Revision 1, Document Version 8.20. Supersedes ERMS 550239."* ERMS 555358, Carlsbad, New Mexico, Sandia National Laboratories.

Xiong, Y.-L., 2011b. *"Release of EQ3/6 Database DATA0.FM1."* ERMS 555152, Carlsbad, NM: Sandia National Laboratories.

Xiong, Y.-L., 2013a. "An aqueous thermodynamic model for solubility of potassium ferrate in alkaline solutions to high ionic strengths at 283.15 K to 333.15 K." *Journal of Solution Chemistry* 42:1393-1403.

Xiong, Y.-L., 2013b. "A thermodynamic model for silica and aluminum in alkaline solutions with high ionic strength at elevated temperatures up to 100°C: Applications to zeolites." *American Mineralogist* 98:141-153.

Xiong, Yongliang, 2014. "A Pitzer model for the Na-Al $(OH)_4$-Cl-OH system and solubility of boehmite (AlOOH) to high ionic strength and to 250°C." *Chemical Geology* 373:37-49.

Xiong, Y.-L., 2015. "Experimental determination of lead carbonate solubility at high ionic strengths: a Pitzer model description." *Monatshefte für Chemie-Chemical Monthly*146:1433-1443.

Xiong, Y.-L., Nowak, J., Brush, L. H., Ismail, A. E., Long, J. J., 2010. "Establishment of Uncertainty Ranges and Probability Distributions of Actinide Solubilities for Performance Assessment in the Waste Isolation Pilot Plant." *Materials Res. Soc. Symp. Proc.* 1265, 15-20, Materials Research Society.

Xiong, Y.-L., Kirkes, L., and Westfall, T., 2013a. "Experimental Determination of Solubilities of Sodium Tetraborate (Borax) in NaCl Solutions, and A Thermodynamic Model for the Na-$B(OH)_3$-Cl-SO_4 System to High Ionic Strengths at 25°C." *American Mineralogist* 98:2030-2036.

Xiong, Y.-L., Kirkes, L., Westfall, T., and Roselle, R., 2013b. "Experimental determination of solubilities of lead oxalate (PbC_2O_4(cr)) in a NaCl medium to high ionic strengths, and the importance of lead oxalate in low temperature environments." *Chemical Geology* 342:128-137.

ABOUT THE EDITOR

Yongliang Xiong, PhD
Sandia National Laboratories (SNL)
Carlsbad Programs Group
Carlsbad, NM, US
yxiong@sandia.gov

INDEX

A

absorption spectroscopy, 195, 196, 199
acid, 4, 21, 27, 28, 29, 160, 164, 166, 167, 182, 185, 198, 199
acidic, 33, 34, 37, 212, 220
actinide, vi, 39, 67, 100, 101, 129, 130, 229, 230, 238, 239, 240
activation energy, 147, 149
additives, 155, 172, 199
adsorption, 128, 129, 140, 205
alcohols, 157
algae, 160, 162
alkaline, vi, 37, 229, 230, 231, 236, 237, 238, 240
$Am(OH)_3(s)$, vi, ix, 229, 231, 232, 233, 235, 236
americium, 129, 238
antimony, 162, 172, 176
aqueous solutions, 2, 45, 67, 157, 166, 212, 220, 239
argon, 4, 50, 193, 201
aromatics, 161, 187
Arrhenius equation, 149
arsenic, 161, 172, 175, 197, 198
atomic emission spectrometry, 193, 201
atoms, 74, 75, 140, 208, 209, 210, 214, 216, 218, 222

B

barium, 162, 173, 175
benzene, 161, 222
binding energies, 26
biocompatibility, 135, 136, 146
biodegradable materials, 134
biodiesel, 175, 202
biological systems, 134
biomarkers, 163, 193
biomaterials, 134, 135
biomedical applications, viii, 134, 135, 146, 154
blends, 136, 137, 148, 149, 154, 155
bonding, 33, 39, 40, 147, 149, 150
brucite, v, vii, 47, 48, 49, 50, 52, 53, 56, 57, 58, 59, 60, 61, 62, 65, 67, 68, 98, 104, 106, 115, 124
brucite chimneys, 48

C

Ca^{2+}, 13, 19, 101, 102, 126

cadmium, 162, 171, 174, 175, 196, 197, 198, 203
calcium, 13, 79, 161, 162, 170, 173, 175
carbon, 27, 47, 49, 140, 164, 166, 186, 230
carbon atoms, 140
carbon dioxide, 47, 49
catalyst, 84, 163
cellulose, 137, 155
cerium, 162, 179, 181, 185, 204
cesium, 105, 119, 120, 124, 125, 128
chain molecules, 137, 154
charge density, 136, 142, 145, 148
chemical, 4, 6, 29, 43, 47, 77, 102, 105, 129, 133, 135, 146, 149, 160, 161, 163, 172, 179, 186, 194, 197, 198, 212, 213, 214, 224
chemical cross-links, 149
chemical properties, 133, 179
chladni modes, 84, 86
chlorine, 139, 140
chlorophyll, 163, 191
chromatographic technique, 162
chromium, 162, 170, 173, 175, 198, 199, 216
classification, 178, 192
cobalt, 4, 162, 171, 173, 175, 197
colloidal transport, 99
colloids, vi, viii, 97, 100, 101, 102, 103, 104, 105, 109, 110, 112, 115, 119, 120, 124, 125, 126, 127, 128, 129, 130, 131
combustion, 160, 164, 166, 168, 185, 189, 195, 200, 202
composites, 135, 137, 156
composition, 10, 19, 26, 27, 32, 42, 81, 140, 141, 142, 145, 146, 148, 149, 150, 187, 219
compounds, 2, 24, 26, 44, 75, 136, 143, 162, 163, 187, 196, 197, 214
contamination, 67, 166, 168, 186
cooling, 6, 27, 167
copper, 161, 162, 170, 173, 175, 193, 196, 198, 199, 216
correction factors, 48, 50, 51, 53, 54, 55, 65
Coulomb interaction, 137, 143, 144, 145
crude oil, viii, 159, 160, 161, 162, 163, 164, 165, 166, 167, 168, 170, 173, 175, 177, 178, 179, 180, 181, 182, 185, 187, 188, 189, 190, 191, 192, 193, 194, 196, 197, 198, 199, 200, 201, 202, 203, 204
crystal structure, 84, 215, 225
crystalline, 47, 49, 74, 84
crystallization, 2, 27
crystals, 23, 69, 96
CVG-AAS, 160, 172, 186

D

decomposition, 26, 80
dehydration, 82, 85
Delta, 176, 193, 202
density fluctuations, 74
Department of Energy, xi, 49, 66, 88, 98, 127, 237
deposition, 2, 27, 164, 166
deposits, vii, 2, 31, 37, 70, 71, 79
depth, 2, 75, 137
derivatives, 154, 174, 191, 222, 223
detectable, 178, 181, 184
detection, 168, 186, 220
determination, v, vi, viii, 34, 40, 41, 42, 47, 68, 69, 72, 94, 95, 97, 105, 130, 138, 141, 157, 159, 160, 163, 164, 165, 167, 168, 169, 170, 171, 172, 173, 176, 179, 185, 186, 187, 188, 189, 190, 191, 192, 193, 194, 195, 196, 197, 198, 199, 200, 201, 202, 203, 204, 240
dielectric constant, 7
different ratios, 148
diffraction, 72, 74, 77
digestion, 160, 164, 166, 167, 182, 185, 189, 194, 195, 198, 199, 202
direct introduction, 159, 160, 164, 165, 185
distillation, 160, 174, 190, 200, 203

distilled water, 4, 10, 80
distribution, 12, 30, 31, 33, 70, 157, 179, 181, 185, 205, 210, 213, 216
DMF, 139, 142, 147, 148, 217
DOI, 38, 39, 40, 41, 42, 43, 44, 45, 46
donors, 209, 210, 215, 217, 218, 219
drug delivery, 134, 156
dynamic viscosity, 147, 148

E

electrolyte, 138, 156
electromagnetic, 208
electron, 52, 53, 72, 210, 211, 214, 216, 227
electron paramagnetic resonance (EPR), viii, 208, 210, 214, 215, 216, 218, 220, 224, 226
electrons, 208, 210, 213, 222, 223
emission, 51, 103, 105, 160, 164, 173, 186, 190, 200, 201
emulsions, 165, 187, 196, 201
energy, viii, 70, 75, 76, 80, 84, 166, 170, 187, 208, 209, 227
engineered barrier, viii, 47, 49, 98, 99, 102
environment, viii, 48, 49, 69, 75, 88, 95, 129, 159, 162, 163, 178, 179, 180, 185, 203, 216, 226, 230, 231, 240
environmental chemistry, 99, 130, 192
environmental conditions, 134
equilibrium, ix, 3, 6, 11, 14, 15, 27, 28, 29, 31, 35, 36, 45, 50, 55, 62, 63, 64, 66, 68, 144, 164, 217, 226, 230, 232, 234, 239
ET-AAS (GF-AAS), 160
europium, 162, 179, 180, 181, 185
experimental condition, 64
experimental design, 199
extinction, 164, 166, 185, 186
extraction, 187

F

field theory, 208
films, viii, 133, 150
filters, 80, 99, 100, 103, 104, 105, 175
filtration, 99, 103, 104, 135
flame, 160, 170, 186, 193, 196
flame atomic absorption spectrometry (FAAS), 160, 170, 173, 186, 193, 196
flow curves, 148
fluid, viii, 1, 2, 3, 5, 6, 9, 10, 11, 16, 21, 32, 34, 35, 37, 40, 42, 43, 46, 69, 71, 72, 73, 76, 77, 78, 79, 82, 84, 85, 86, 88, 89, 95, 187
fluid inclusions, viii, 40, 69, 71, 72, 73, 76, 78, 79, 82, 84, 85, 86, 89, 92, 94, 95
fluorine, 13, 27, 31, 38
formation, vii, 2, 7, 14, 15, 24, 26, 27, 28, 29, 30, 31, 32, 33, 37, 38, 41, 42, 43, 48, 49, 50, 65, 67, 72, 73, 76, 98, 99, 101, 128, 144, 145, 146, 159, 163, 185, 219, 227, 234
formula, 24, 27, 36, 87, 220
fragments, 100, 101, 102
free energy, 29
fuel cell, 135, 155
fuel oils, vi, viii, 159, 160, 168, 170, 172, 174, 176, 188, 194
functionalized polysulfones, viii, 133, 134, 135, 151, 152, 154

G

geology, 70, 193
geometry, 211, 212, 214, 215, 216, 217, 219, 224
germanium, 162
grain boundaries, viii, 69, 71, 73, 76, 77, 78, 79, 82, 84, 85, 88
graphite, 160, 170, 186, 193, 197, 198, 199
gravity, 188

groundwater, 67, 129

H

hafnium, 46, 162
Hanford site, 230, 234, 235, 236, 238
heavy metals, 97, 163, 201
heavy oil, 168, 173, 198, 200
high density polyethylene, 50
high level nuclear waste, 48
high level tank waste, 230, 234, 236
holmium, 162, 179, 205
homogeneity, 165, 168
homopolymers, 152
hydrocarbons, 160, 161
hydrogen, 4, 33, 34, 40, 48, 50, 51, 55, 65, 73, 75, 84, 144, 147, 149, 150, 161
hydrogen atoms, 75, 84
hydrogen bonds, 144
hydrolysis, 32, 33, 38, 42, 44, 48, 67, 68, 100, 208, 226, 227, 230, 237
hydrothermal synthesis, 4, 19
hydrothermal system, vii, 26, 38, 41, 67
hydroxide, 44, 68, 93, 101, 102, 131, 221
hyperalkaline, vi, viii, ix, 48, 49, 65, 67, 229, 231, 237

I

ICP-MS, 4, 73, 95, 104, 105, 160, 162, 164, 166, 167, 169, 170, 173, 174, 186, 188, 189, 190, 193, 194, 195, 200, 202, 203
ICP-OES, 160, 164, 169, 170, 173, 175, 186, 193, 195, 200
in transition, 208, 209, 217
information processing, 134
infrared spectroscopy, 188
interactions, 68, 84, 90, 134, 136, 137, 138, 140, 142, 143, 144, 145, 146, 147, 148, 149, 150, 152, 156, 157, 192, 237
interface, 71, 84, 154, 221
intergranular water, viii, 69, 77, 87
intermolecular interactions, 144, 149, 157
intrinsic viscosity, 138, 141, 142
ionic polymers, 136
ions, 7, 24, 33, 44, 102, 126, 175, 186, 209, 211, 213, 220, 222
iron, 105, 161, 162, 170, 173, 175, 196, 199
isolation, 47, 48, 65, 68, 70, 98, 100, 130, 131, 161, 162
isomers, 216, 217
isotope, 130, 175, 189, 203, 204, 211, 212, 216

K

K^+, 101, 102, 126, 229, 231, 233, 236
KOH, vi, ix, 229, 232, 233

L

lanthanum, 162, 176, 179
ligand, 208, 212, 214, 217, 218, 219, 222, 223
light, 23, 70, 73, 168, 172, 175, 180, 194, 202
liquid chromatography, 162, 191
liquid phase, 32
lithium, 27, 93, 161, 175
lubricating oil, 171, 174, 189, 196, 201
lutetium, 162, 179, 181

M

macromolecular chains, 144
macromolecules, 136, 143, 144
magnesium, viii, 49, 51, 53, 55, 67, 68, 79, 98, 99, 101, 102, 104, 105, 106, 109, 112, 115, 119, 125, 130, 162, 170, 173, 175

mass, 6, 15, 21, 73, 80, 84, 104, 105, 160, 162, 163, 164, 169, 170, 173, 174, 179, 182, 186, 189, 190, 191, 202, 203, 204
mass spectrometry, 160, 162, 164, 173, 186, 189, 190, 191, 202, 203, 204
melt, 2, 21, 27, 31, 32, 35, 36, 40, 42, 43
melting, 27
melts, 3, 26, 30, 34, 35, 36, 42
membranes, 133, 135, 136, 146, 147, 155, 156
mercury, 162, 172, 176, 189, 193, 199
metal complexes, 208, 209, 217
metal ion, viii, 99, 161, 187, 207, 208, 209, 216, 218, 219, 220, 221, 222, 223, 226
metals, 2, 28, 38, 43, 161, 163, 185, 189, 191, 193, 194, 196, 200, 201, 202, 204, 211, 220, 221, 222
$Mg(OH)_2$, vii, 47, 48, 49, 50, 52, 53, 54, 66, 98, 104
Mg^{2+}, 48, 51, 54, 63, 101, 102, 126
MgO, viii, 47, 49, 67, 68, 98, 102, 103, 105, 106, 107, 108, 109, 110, 111, 112, 113, 114, 115, 116, 117, 118, 119, 120, 121, 122, 123, 124, 125, 126, 127, 128, 129, 130
microemulsion, 176, 189, 196, 199, 204
microlite, vii, 1, 2, 4, 5, 7, 8, 9, 10, 11, 12, 13, 15, 29, 30, 35, 43
microscope, 19, 23, 52, 53, 72
mineral, vi, viii, 3, 4, 6, 7, 46, 49, 72, 81, 92, 96, 97, 98, 100, 101, 102, 103, 104, 105, 109, 110, 112, 115, 119, 120, 124, 125, 126, 127, 128, 129, 163
mineralization, 37, 159, 160, 164, 165, 166, 185, 188
mixing, 142, 145, 147, 148, 151, 233
moisture, 70, 72, 73, 78, 82, 83, 84, 85, 86, 87, 88
molecular dynamics, 40
molecular mass, 225
molecular structure, 146, 157
molecular weight, 99, 103, 136, 160
molecules, viii, 24, 29, 33, 69, 84, 86, 88, 93, 141, 142, 145, 146, 210, 212, 213, 219, 220
molybdenum, 162, 173, 175
morphology, 71, 75, 134, 146, 151
multiple trace elements, 159, 160

N

Na^+, 13, 19, 51, 101, 102, 126, 229, 231, 233, 236
Na2SO4, 103, 104, 235
NaCl, v, vii, 3, 30, 45, 47, 48, 50, 51, 52, 53, 54, 55, 56, 57, 58, 59, 60, 61, 62, 63, 64, 65, 67, 79, 81, 84, 88, 94, 102, 103, 104, 106, 125, 128, 240
nebulizer, 164, 173, 174, 190
neutral, 35, 136, 139, 140, 142, 144, 145, 149, 212, 221
neutron, viii, 69, 72, 73, 74, 75, 78, 80, 83, 86, 89, 90, 91, 92, 93, 94, 95, 96, 129, 192
nickel, 3, 161, 163, 170, 173, 175, 178, 179, 180, 187, 192, 193, 196, 197, 198, 199, 201, 203, 204
niobium, v, 1, 2, 7, 16, 21, 28, 29, 30, 35, 37, 38, 41, 42, 45, 46, 162
nitrogen, 80, 161, 163, 218, 219
nitrogen gas, 80
nitrous oxide, 196
non-metals, 194
nuclear magnetic resonance (NMR), viii, 72, 73, 188, 208, 210, 212, 213, 220, 224
nuclear waste, vi, viii, 47, 48, 65, 67, 68, 70, 71, 72, 75, 76, 79, 90, 91, 92, 94, 95, 97, 98, 99, 100, 127, 130, 131, 230
nuclear waste management, vi, 48, 90, 94, 97, 99, 230
nucleus, 70, 211, 212, 213

O

oil, 159, 160, 161, 162, 163, 164, 165, 166, 168, 169, 170, 172, 174, 177, 178, 179, 180, 181, 185, 186, 187, 189, 192, 195, 196, 198, 199, 200, 201
oil samples, 160, 164, 166, 168, 169, 170, 172, 174, 175, 177, 178, 179, 180, 181, 186, 198
oil spill, 163, 185
organ, 165, 171, 196
organic compounds, 160
organic matter, 128, 159, 163, 185
organic solvents, 165
oxalate, 215, 217, 240
oxidation, 40, 44, 179, 207, 208, 209, 210, 211, 212, 222, 223, 227, 233, 234
oxygen, 4, 42, 130, 161, 163, 201, 209

P

peptides, 218, 226
permeability, 71, 100
petroleum, 67, 160, 172, 176, 187, 188, 190, 191, 192, 193, 194, 196, 197, 198, 199, 201, 203
pH, viii, 28, 29, 48, 50, 51, 56, 57, 58, 59, 60, 61, 62, 65, 135, 137, 178, 212, 213, 214, 219, 220, 230, 238
phosphorus, 161, 174, 175
physicochemical properties, 134, 136
Pitzer model, ix, 229, 230, 231, 232, 236, 240
platinum, 3, 23, 162
poly(methyl methacrylate), 157
poly(vinyl chloride), 154
polyelectrolyte effect, 134, 136, 149
polymer, 135, 136, 137, 138, 139, 140, 141, 142, 143, 144, 145, 146, 148, 149, 150, 153, 155, 156
polymer blends, 148, 155
polymer chain, 142, 145, 149
polymer materials, 143, 146
polymer solutions, 141, 156
polymeric materials, 134, 143
polynuclear complexes, 220
polyvinyl alcohol, 137, 140
porphyrins, 161, 191
potassium, 27, 79, 161, 162, 174, 175, 225, 240
preparation, iv, 50, 77, 78, 152, 159, 160, 164, 166, 185, 188, 194, 196, 198, 199, 201
protons, 219
PSF-DMBA, 140, 141, 143, 144, 145, 147, 148, 149, 150, 151
PSF-DMEtA1, 139, 140, 148, 150
PVA, 137, 140, 141, 144, 145, 147, 148, 150, 151
pyrochlore, vii, 2, 4, 5, 7, 15, 16, 17, 18, 19, 21, 23, 29, 30, 35, 43

Q

QQQ-ICP-MS, 160, 175
quartz, 40, 71, 79, 82, 168
quaternary ammonium, 135

R

radioactive waste, 66, 98, 128, 239
radionuclides, vii, viii, 97, 237, 238, 239
rare earth elements, 68, 129, 175, 179, 202
reciprocal temperature, 64, 65
residue, 71, 78, 80, 81, 82, 83, 85, 86, 173, 188, 200
resins, 161, 187
resistance, 135, 136
rheology, 69, 76, 153, 155
room temperature, 27, 49, 68, 81, 82, 83, 86, 131
rotor-based microwave, 160

rubidium, 162, 175

S

salt, v, vi, viii, 20, 21, 26, 32, 33, 38, 45, 47, 48, 49, 65, 66, 67, 68, 69, 70, 71, 72, 73, 75, 76, 77, 78, 79, 80, 81, 82, 83, 85, 88, 89, 90, 91, 92, 93, 94, 95, 97, 98, 100, 101, 102, 103, 104, 127, 130, 131, 152, 153, 196, 212, 230, 236, 237
salt formation, vi, viii, 47, 48, 49, 65, 68, 70, 76, 78, 88, 98, 99, 100, 101, 102, 127, 131, 230, 236
sample preparation, 77, 78, 159, 160, 164, 165, 166, 185, 188, 194, 196, 198, 201
Savannah River site, 230, 234, 236, 238
scattering, viii, 69, 72, 73, 74, 75, 77, 83, 84, 86, 92, 93, 94
scattering intensity, 73, 84
science, 135, 136, 137, 143, 188
sedimentation, 162
sediments, 128, 191, 192
selenium, 161, 171, 175
sensitivity, 30, 75, 169, 170, 172, 186
side chain, 135, 136, 140, 141, 153
silica, 127, 128, 240
silicon, 161, 173
silver, 162, 175
SiO2, 27, 38, 79, 100, 105
smart materials, 134, 146
SO42, 236
sodium, 3, 12, 13, 21, 24, 26, 32, 35, 66, 162, 171, 173, 176, 199, 201
software, 7, 16, 27, 29, 38, 40, 45
solid phase, 4, 35
solid solutions, 35, 129
solid state, 217, 218
solid surfaces, 92
solubility, vii, ix, 1, 2, 3, 5, 6, 7, 8, 9, 10, 11, 13, 14, 15, 16, 17, 20, 21, 22, 23, 24, 27, 28, 29, 30, 31, 33, 35, 36, 37, 41, 42, 43, 45, 46, 47, 48, 49, 50, 52, 55, 64, 65, 68, 70, 80, 130, 135, 140, 149, 156, 229, 230, 231, 232, 236, 237, 238, 239, 240
solution, vii, viii, 2, 3, 4, 5, 7, 8, 9, 10, 11, 12, 13, 14, 15, 17, 19, 21, 22, 23, 24, 25, 27, 29, 30, 31, 32, 33, 35, 36, 37, 41, 42, 50, 51, 52, 53, 65, 99, 100, 101, 103, 104, 106, 107, 108, 109, 110, 111, 112, 113, 114, 115, 116, 117, 118, 119, 120, 121, 122, 123, 124, 125, 126, 128, 134, 136, 137, 138, 139, 140, 141, 144, 145, 146, 149, 150, 152, 153, 154, 155, 157, 165, 175, 190, 203, 207, 208, 210, 213, 214, 215, 217, 218, 219, 221, 223, 237, 239
solution properties, 134, 150, 153, 154
solvent molecules, 144
solvents, 137, 150, 152, 153, 154, 155, 157, 188, 208, 217, 223
sorption, 100
speciation, viii, 8, 18, 27, 41, 45, 130, 212, 213, 223, 230, 239
species, 2, 7, 12, 13, 15, 16, 18, 21, 27, 28, 29, 30, 31, 33, 35, 40, 44, 100, 166, 171, 172, 186, 191, 193, 208, 209, 210, 212, 213, 218, 219, 221, 231, 233
specific, 2, 48, 50, 68, 71, 78, 84, 134, 136, 139, 141, 143, 145, 146, 148, 150, 159, 163, 178, 185, 207, 208, 209, 220, 221, 223
specific ion interaction theory (SIT), 48, 63, 68
spectrophotometry, 196, 199, 218
spectroscopic techniques, 208
spectroscopy, viii, 45, 75, 80, 190, 195, 196, 200, 208, 210, 215, 216, 218, 220
SRC microwave, 160, 167, 168, 182
steel, 168, 230
stimuli-sensitive, 135
strontium, 162, 170, 175
sulfate, 79, 204

sulfur, 173, 175, 178, 179, 189, 193, 200, 201
surfactant, 165, 198
symmetry, 214, 215
synergistic effect, 137
synthesis, 4, 19, 42, 143
synthetic methods, 143
synthetic polymers, 136

T

tantalum, v, 1, 2, 7, 8, 10, 11, 12, 14, 28, 29, 30, 35, 37, 38, 45, 46, 162, 178
temperature, vii, 7, 11, 29, 30, 31, 32, 38, 42, 45, 48, 50, 51, 53, 64, 65, 67, 69, 70, 75, 78, 80, 82, 83, 84, 85, 86, 88, 142, 153, 154, 157, 165, 166, 167, 182, 185, 191, 214, 217, 231, 240
thermal properties, 152
thermodynamic calculations, 9, 13, 16, 30, 102
thermodynamic modeling, v, vii, 1, 2, 230, 239
thermodynamic properties, 7, 28, 29, 30, 40, 44, 68
tissue engineering, 134, 152
titanium, 162, 174, 175
toluene, 190, 217
trace elements, vi, viii, 129, 159, 160, 163, 164, 165, 170, 174, 178, 185, 186, 189, 190, 191, 192, 194, 195, 198, 203
transition metal, 207, 208, 210, 211, 216, 220, 223
transition metal ions, 207, 208, 210, 211, 216, 220, 223

U

uraninite, vii, 1, 2, 4, 5, 19, 21, 23, 24, 26, 27, 31, 35, 41, 43, 46
uranium, v, 1, 2, 7, 19, 21, 24, 26, 27, 28, 30, 31, 35, 37, 38, 39, 40, 41, 42, 43, 44, 162, 175, 237, 238
UV-Vis spectroscopy, viii, 208, 218

V

vanadium, vi, viii, 81, 161, 163, 170, 173, 175, 177, 178, 179, 180, 187, 191, 192, 193, 194, 196, 197, 198, 199, 200, 201, 203, 204, 207, 208, 210, 212, 213, 214, 220, 224, 225, 227
vapor, 32, 33, 66, 160, 172, 176, 186, 189, 199
viscoelastic properties, 148
viscosity, 136, 137, 138, 139, 140, 142, 147, 148, 150, 154, 155, 157, 164, 165, 166, 169, 185, 214

W

waste, viii, 47, 48, 49, 65, 67, 68, 70, 71, 72, 75, 76, 79, 98, 100, 127, 130, 131, 174, 201, 230, 234, 235, 236, 237, 238
waste isolation pilot plant (WIPP), 47, 48, 49, 67, 76, 79, 94, 98, 99, 100, 101, 102, 126, 127, 129
waste management, 48, 230
water, v, viii, 3, 4, 6, 7, 8, 10, 16, 23, 27, 28, 29, 30, 33, 34, 35, 37, 39, 40, 42, 50, 51, 63, 64, 68, 69, 70, 71, 73, 74, 75, 76, 77, 78, 79, 80, 82, 83, 84, 85, 86, 87, 88, 91, 92, 93, 94, 95, 96, 135, 139, 140, 142, 147, 148, 162, 163, 165, 171, 178, 185, 196, 198, 210, 215, 219, 220, 233, 237, 239
water permeability, 135
water supplies, 163
water vapor, 68

X

X-ray diffraction, 4, 20, 21, 22, 24, 51

Y

yttrium, 68, 162, 176, 179, 204

Z

zeolites, 71, 84, 85, 92, 240
zinc, 161, 162, 171, 173, 175, 198, 199
zirconium, 46, 162